PREVENTION OF POST-HARVEST FOOD LOSSES:
A TRAINING MANUAL

FAO Training Series   *No. 10*

Prevention of post-harvest food losses: a training manual

FOOD AND AGRICULTURE ORGANIZATION OF THE UNITED NATIONS
Rome, 1985

Acknowledgements

Acknowledgement is due to D.J. Greig and M. Reeves, FAO Consultants. Mr Greig compiled the document; Mr Reeves reviewed and edited the text. Thanks are also due to technical officers of FAO who contributed to the preparation of this manual.

The designations employed and the presentation of material in this publication do not imply the expression of any opinion whatsoever by the Food and Agriculture Organization of the United Nations concerning the legal status of any country, territory, city or area or of its authorities, or concerning the delimitation of its frontiers or boundaries. The views expressed are those of the authors.

P-00
ISBN 92-5-102209-7

The copyright in this book is vested in the Food and Agriculture Organization of the United Nations. The book may not be reproduced, in whole or in part, by any method or process, without written permission from the copyright holder. Applications for such permission, with a statement of the purpose and extent of the reproduction desired, should be addressed to the Director, Publications Division, Food and Agriculture Organization of the United Nations, Via delle Terme di Caracalla, 00100 Rome, Italy.

© FAO 1985

Printed in Italy

Foreword

Since 1977, FAO has given priority to the prevention of post-harvest losses, particularly through action to reduce losses at the farm and village level. A serious constraint for developing countries in organizing and implementing post-harvest loss prevention programmes is a shortage of trained national staff.

The FAO Action Programme for the Prevention of Food Losses has undertaken a training programme during which several workshops were organized to train African technical officers to reduce post-harvest losses. The workshops covered various aspects of crop storage and processing, including storage pests and their control, loss assessment, drying, storage, grain processing and socio-economic implications.

The material in this manual has been tested during training courses, and has also been reviewed on the basis of experience gained. It is now being published as part of the FAO Training Series.

I trust that this manual will help to provide practical training for those responsible for the prevention of post-harvest food losses in developing countries.

D.F.R. Bommer
Assistant Director-General
Agriculture Department

Preface

This manual presents material from a wide range of disciplines associated with the prevention of food losses in, particularly, cereals, pulses, roots and tubers. It is directed at field staff, project supervisors and extension personnel involved in food-loss prevention programmes.

The manual should serve as a single-volume reference work on food-loss prevention during storage. While the approach is technical in the sections on drying, processing and loss assessment, the economic and social aspects of food losses are also considered.

It is hoped that participants in training workshops will find that the manual meets their basic needs, and that it will be supplemented by detailed worksheets covering special topics (especially for practical work), and by handouts on subjects of particular local importance.

Contents

Foreword		v
Preface		vi
1. Introduction		**1**
1.1	Definitions	1
1.2	Storage	2
1.3	Storage requirements	4
1.4	Agents causing deterioration of stored produce	4
1.5	Controlling the agents causing deterioration	5
2. Measurement		**9**
2.1	Introduction	9
2.2	Units of measurement	9
2.3	Repeatability and precision	10
2.4	Measurement of moisture content	10
2.5	Sampling for loss assessment	12
2.6	Farm- and village-level grain flow pattern	15
3. Storage pests		**17**
3.1	Post-harvest microbiology	17
3.2	Pest biology and identification	19
3.3	Descriptive notes	19
4. Loss assessment		**29**
4.1	Introduction	29
4.2	Definitions	29
4.3	Surveys	29
4.4	Field trials	30
4.5	Validity of survey and trials work	31
4.6	Loss assessment in cereals and pulses	31
5. Pest control		**35**
5.1	General	35
5.2	Losses caused by insects	35
5.3	Sources of infestation	36
5.4	Pest build-up in store	37
5.5	Factors affecting choice of storage method and pest control measures	38
5.6	Types of storage and implications for pest control	38
5.7	Losses caused by rodents	42

6.	**Drying**	**47**
	6.1 Introduction	47
	6.2 Air and water vapour: psychrometry	47
	6.3 Moisture content and relative humidity	52
	6.4 Drying	53
	6.5 Types of driers	55
7.	**Warehouses**	**65**
	7.1 Warehouse construction	65
	7.2 Cost of construction	65
	7.3 Usable volume	66
	7.4 Care of produce in warehouses	66
	7.5 Dunnage	69
	7.6 Stacking of sacks	70
	7.7 Insect control in sacks stored in warehouses	71
8.	**Central storage**	**73**
9.	**Pest control in stored produce**	**75**
	9.1 Introduction	75
	9.2 Pest control techniques	75
	9.3 Chemical control: specific methods	80
	9.4 Toxicity	84
	9.5 Formulations and dosages	85
	9.6 Some insecticides for use with stored products: summary of properties	86
10.	**Crib storage**	**89**
	10.1 Introduction	89
	10.2 Optimal crib design	89
	10.3 Design of improved cribs	90
	10.4 Cost of crib construction	90
11.	**Roots and tuber storage**	**93**
	11.1 Yams	93
12.	**Processing of cereals (other than rice)**	**97**
	12.1 Threshing	97
	12.2 Grading	97
	12.3 Milling	100

13.	**Small-scale rice milling**	**105**
	13.1 Introduction	105
	13.2 Stages in rice processing	105
14.	**Sociological, economic and institutional implications of the prevention of post-harvest food losses**	**115**
	14.1 Economic justification	115
	14.2 Institutional factors	118
	14.3 Implications for labour	120
References		**121**

1. Introduction

1.1 Definitions

It is important to understand the principles of sampling and measurement. Only then may the procedures defined here be adapted to suit local conditions, with reasonable confidence that measurements will be valid. The terms defined in the following list are commonly used in dealing with post-harvest food losses.

Post-harvest. The period between maturity of the crop and the time of its final consumption.
Food loss. Any change in the availability, edibility, wholesomeness or quality of food that reduces its value to humans.
Direct loss. Loss by spillage or consumption by insects, rodents and birds.
Indirect loss. Loss caused by a lowering of quality leading to rejection as food. This type of loss may be locally defined and related to custom.
Losses of crop product. Crop products may be lost from the food chain at any or all of the periods between planting and preparation for immediate consumption. Three general periods have been identified.
(a) Pre-harvest losses occur before the harvesting process begins and may be due to such factors as insects, weeds or diseases afflicting the crop.
(b) Harvest losses occur during the harvesting process and may be due, for example, to shattering and shedding of the grain from the ears to the ground.
(c) Post-harvest losses occur during the post-harvest period.
Post-production losses. Losses consisting of the combined harvest losses and post-harvest losses.

It is always difficult to distinguish clearly between the arbitrarily defined stages from production to consumption. The maturing/drying/processing periods will often overlap during the post-harvest period, as, for example, in the field-drying of maize after it has reached maturity. There is nothing to be gained by defining rigid boundaries and making artificial distinctions between overlapping stages. It may be preferable to relate losses to a process or operation rather than to a definite period.
Food. Those commodities which people normally eat: the weight of wholesome edible material, measured on a moisture-free basis, that would normally be consumed by humans. Inedible portions of the crop, such as stalks, hulls and leaves, are not food. Crops for consumption by animals are not considered food. Post-harvest loss assessments are generally made on the basis of

dry-matter changes. Normally, no attention is paid to nutritional or financial losses.

Grain loss. The loss in weight, occurring over a specified period and expressed on a moisture-free basis, of grain which would otherwise have been available as human food.

Moisture content (mc). The quantity of free water in a specified material. Materials of organic origin are defined for scientific purposes as consisting of dry matter and water. Loss of moisture during drying is not a food loss. Moisture content is expressed either as a decimal ratio or as a percentage in one of two ways.

(a) Wet basis (wb). The moisture content is defined as the ratio of the weight of water to the total weight of dry matter and water. This is the most commonly used method in agriculture.

(b) Dry basis (db). The moisture content is defined as a ratio of the weight of water to the dry-matter weight. This method is normally used in scientific laboratory work.

In agriculture it is traditional to use wet-basis moisture contents. Where moisture contents are expressed without an indication of wb or db, it may be assumed that the moisture contents are on a wet basis.

Decimal numerical values may be transformed from one basis to the other by the following relationships:

$$\text{mcwb} = \frac{\text{mcdb}}{1 + \text{mcdb}}$$

$$\text{mcdb} = \frac{\text{mcwb}}{1 - \text{mcwb}}$$

It should be noted that the values given by these formulas differ for the same sample.

1.2 Storage

Farmers produce crop products. Some of these require some processing before becoming suitable as food for humans. Crop products become available during different short periods of the year, but people wish to consume the food steadily throughout the year. Some form of storage is therefore required.

The storage requirements of crops show wide variation. For durables, such as cereal grains, the requirements are comparatively simple; while for perishable crops, such as fruit or vegetables, the cost of providing long-term storage is very high. Such difficulties may be overcome either by lengthening the production season of the perishables, or by partially or completely processing them into a more concentrated form.

Figure 1.1 *Traditional storage structures*

1.3 Storage requirements

The crop product must be stored so that:
(a) the quality does not deteriorate during the storage period;
(b) the quantity in storage is not unintentionally reduced;
(c) it is secure against pests, diseases and physical loss; and
(d) it is accessible at the time and in the quantity required.

The main crop products which may require storage facilities are:
- durables (cereal grains)
- perishables (fruits and vegetables)
- semi-durables (roots and tubers)

Some processing may be necessary for the perishable products. Special care and specialized structures may be necessary for semi-durables, such as yams and sweet potatoes, before they can be stored successfully. The cost of processing and the cost of storage are important considerations in planning storage strategy.

Durable crop products are relatively easy to store, compared with the other two categories.

1.4 Agents causing deterioration of stored produce

The main agents causing deterioration of stored produce are:
- microorganisms (fungi, bacteria and yeasts)
- insects and mites
- rodents
- birds
- metabolic activity

Fungi. The most important type of microorganism causing or supporting crop deterioration. Although belonging to the plant kingdom, fungi possess no chlorophyll and are therefore unable to manufacture their own food by photosynthesis. Thus, they live on other living bodies as *parasites*; or on inactively alive or dead bodies as *saprophytes*. Parasitic fungi may cause diseases in the host body, while saprophytic fungi degrade or destroy the body on which they feed. Saprophytic fungi are more important in relation to stored durable crops.

Bacteria. Not generally a problem with dry-stored durables. They may, however, invade already damaged portions of the crop product during storage, and multiply.

Insects. Many species of insect are found in stored crop products, but only a few cause damage and loss. Some may even be beneficial because they attack other insect pests. It is important to be able to identify accurately the main

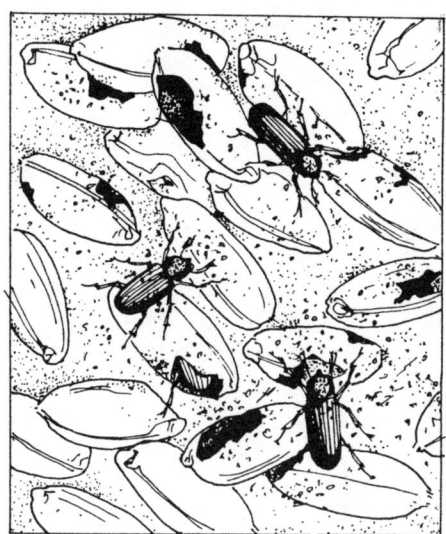

Figure 1.2 *Insect pests*

Figure 1.3 *Rodent damage*

insect species in order to assess their effect on the stored product and to devise the necessary control measures.

Rodents. Rodents prefer not to live in grain stores because there is no drinking-water. Although they can subsist without freely available water, the climate in the store is too dry for them to multiply rapidly unless they can leave the store to find water and return easily. Rodents consume grains and damage sacks and building structures, but they contaminate much greater quantities of grains with urine and droppings than they consume. They are controlled by poisoning and by preventing their access to the stored commodities.

Birds. Like rodents, birds consume some grain but also contaminate a greater quantity with their droppings. Losses caused by birds are avoided by preventing their access to the stored commodities.

Metabolic activity. The crop product is living material, and its normal chemical reactions produce heat and chemical by-products. Heat is also generated by insects, mites and microorganisms which, if present in large numbers, may lead to a significant rise in the temperature of the stored product.

1.5 Controlling the agents causing deterioration

The agents causing deterioration (with the exception of a few anaerobic species) require moisture, oxygen and an equable temperature in order to multiply and thereby damage the product.

Such agents are controlled by keeping one or more of these factors at levels which prevent (or at least deter) their growth or by measures such as the application of insecticides, or fungicides (e.g. propionic acid).

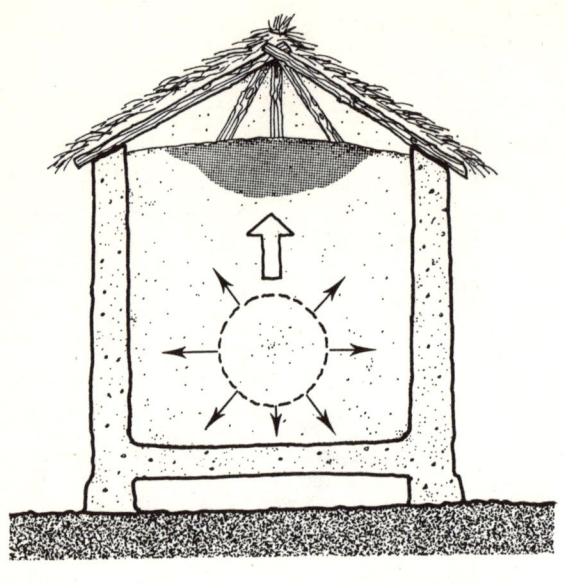

Figure 1.4 *Metabolic activity*

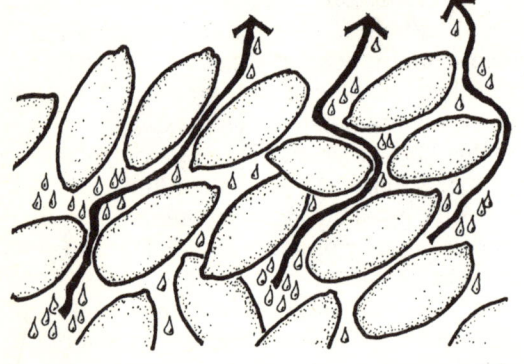

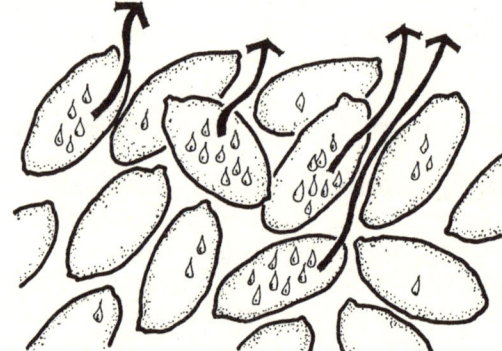

Figure 1.5 *Reduction of moisture in cereals*

1.5.1 **Reduction of moisture content.** The rate of metabolic activity is significantly reduced in most cereals if the grain moisture content is reduced to 14 percent; below 8 percent, metabolic activity practically ceases. Drying is therefore a standard treatment for wet cereal crops before storage. Drying requires energy to evaporate the moisture, and air movement to remove the resultant water vapour. The energy may be derived from burning fossil fuel or wood, or from solar energy, as in sun-drying. It may also be derived from ambient air that is not fully saturated with vapour (as in the crib drying of ear maize). Air movement may arise through convection currents caused by very small temperature differences, by a general air movement such as wind or a breeze, or by artificial means such as a fan. Drying processes are well documented and results can be reliably predicted.

1.5.2 **Reducing oxygen.** Bulk grain may be stored in airtight containers to exclude oxygen. If the grain is wet (17-20 percent mc), metabolic activity soon exhausts the initial oxygen supply and the grain will not deteriorate in feeding quality. The germ, however, is destroyed and anaerobic fermentation may lead to unacceptable taints. Such grain is only used for animal feeding. If the grain is dry (12-13 percent mc), it may be stored for several years, with careful management. In controlled (modified)-atmosphere storage, nitrogen or carbon dioxide is often used to replace the original air when the container is first loaded.

1.5.3 **Controlling temperature.** Levels of insect activity and general metabolic activity rise with increasing temperatures up to 42°C. The maintenance of low temperatures in the bulk grain mass by using modern refrigeration techniques has been used successfully to control deterioration and maintain viability of the stored grain. The method is used in specialized fields such as seed storage and storage of grain for brewing. Equipment and running costs are high.

1.5.4 **Chemical control.** The bulk grain is treated with an organic acid or with gaseous ammonia. This sterilizes the grain and kills the germ, and generally leaves an odour disagreeable to humans in the grain, which is then used for stock feed. Insecticide and fumigant treatments may also be considered as chemical control methods.

2. Measurement

2.1 Introduction

Observation by an experienced person can often identify a problem and suggest a possible solution. It is extremely difficult, however, for even an experienced observer to predict precisely the value of the proposed solution in terms of cost and potential benefits. Measurements, made correctly, allow the problem and possible solution to be quantified. They also enable other workers in the field to benefit from the results.

2.2 Units of measurement

Primary measurements are *mass, length, temperature* and *time*; all other units are derived from these.

Derived units are those used in everyday measurements and calculations. The more important of these included in this manual are:

Parameter	Unit	Symbol	Other units
Weight	kilogram	kg	tonne (1 000 kg) quintal (100 kg)
Time	second	s	minute (min) hour (hr)
Distance	metre	m	kilometre (km)
Temperature	degree Celsius	°C	degree Fahrenheit (°F)
Area	square metre	m^2	hectare (ha)
Volume	cubic metre	m^3	litre (l)
Density	kilograms/cubic metre	kg/m^3	grams/millilitre (g/ml)
Force	newton	N	
Pressure	pascal	Pa	newton/square metre (N/m^2)

2.3 Repeatability and precision

The degree of precision required in measurement depends on the general magnitude of the measurement and its purpose. To refer to a 100-km car journey in terms of the nearest metre would be pointless; yet units of 1 mm would be too large to describe the size of a grain mite. Errors in measurement arise from the imprecision of instruments and from the repeatability with which they can indicate the same measurements.

When making general measurements or calculations, it is usually sufficient to use four significant figures and to round off the answer to three significant figures; this should be remembered when using electronic calculators which display ten digits!

2.4 Measurement of moisture content

The moisture content of stored crop products is probably the most important single factor involved in providing safe storage. The moisture content of grain is particularly critical. Safe moisture levels for different types of storage cannot be specified because of the large number of other factors involved, such as: grain type and variety; degree of contamination with foreign matter; degree of damage; quantity in store; and provision for aeration. The wetter the grain, however, the greater the risk of loss.

It is therefore important to know the moisture content of the grain when it enters the store and how this changes during storage. In working stores (as opposed to experimental stores) it is not normally necessary to know moisture content with great accuracy; this is difficult anyway because of the problems of sampling. The store manager, however, must have some idea of the moisture content — to within 1 percent is usually sufficient — during storage.

Methods of measuring moisture content fall into two categories: direct measurements and indirect measurements.

Direct measurements of moisture content. For direct measurements the sample is divided into subsamples and each subsample is treated in succession. The moisture content is determined either by weighing a subsample, then removing the water and reweighing the dried sample (the difference in weight being equal to the water initially present); or by collecting and weighing the water given off. The most common method is the first; the water is removed from the sample by heating it in an oven under controlled conditions. The moisture contents of the subsamples are then averaged to give the moisture content of the original sample.

Oven-drying method. The material is weighed, dried in an oven at a specified temperature for a specified time and reweighed. The loss in weight is assumed to be the water in the original material. To avoid mistakes, a standard form is used to record the measured weights.

Accuracy of weighing. Weighing should be accurate to 1 part in 1 000 and the moisture content of subsamples expressed to three significant figures. The size of the subsample required depends on the type of weighing device, the oven space and sample containers available. A typical sample size would be 50 g of grain in a 75-mm diameter × 20-mm deep tin, weighed to 0.01 g.

The drying oven. The drying oven must have fan-assisted ventilation, and its temperature must be adjustable and controllable to ± 1°C over the range 95-135°C.

Time and temperature of drying. Many combinations are used, depending on the environment and whether the grain is whole or ground. The four most commonly used combinations are as follows:
(a) 2 hours at 130°C (ground grain)
(b) 16 hours at 105°C (ground grain)
(c) 72 hours at 100°C (whole grain)
(d) 16 hours at 130°C (whole grain)

Methods (c) and (d) are generally used with whole grain because the process of grinding the grain sample may release moisture before the sample is weighed, thus leading to inaccuracies.

If the grain has more than 25-percent mcwb, two-stage drying is recommended: the first stage at 95°C for the time required to reduce the moisture level to about 18-percent mcwb, and the second for 16 hours at 130°C.

Indirect measurements of moisture content. Indirect methods for determining moisture content measure a property of the grain that is itself related to moisture content. The two most common properties measured are the electrical resistance and the dielectric constant of a grain sample which has been loaded into a measuring cell according to a standard procedure. This ensures that uniform quantities of grain are used and that uniform pressure is exerted on the sample.

The instruments designed to use with these methods present the moisture content as a reading on a scale, or can be used with a conversion scale. On some instruments a correction is necessary to compensate for temperatures outside the calibration range.

There are many commercially marketed meters using these properties. It is essential that the manufacturer's instructions be followed precisely and the instruments recalibrated against oven-dried samples at least once every year.

Another property of grain used for estimating the moisture content is the equilibrium humidity of the hair surrounding the grains. The hair hygrometer responds to relative humidity changes and can be calibrated in terms of grain moisture content. This method is less accurate than the others, and several hours may be necessary for the instrument to reach equilibrium.

The salt test. This is a very simple method for testing the suitability of grain for storage. Dry common salt (non-iodized) is mixed with the grain sample in a glass jar and shaken. The equilibrium relative humidity of dry salt is 75 percent at ambient temperatures. The equilibrium moisture content of grain at 75-percent relative humidity is about 15 percent. So, if the salt in the grain sample adheres to the walls of the glass, it has absorbed moisture from the air which must therefore have been at a relative humidity greater than 75 percent. This means that the grain had a moisture content greater than 15-percent mcwb, and is unsuitable for storage in bulk. The method is not precise, but it costs little and is simple to carry out.

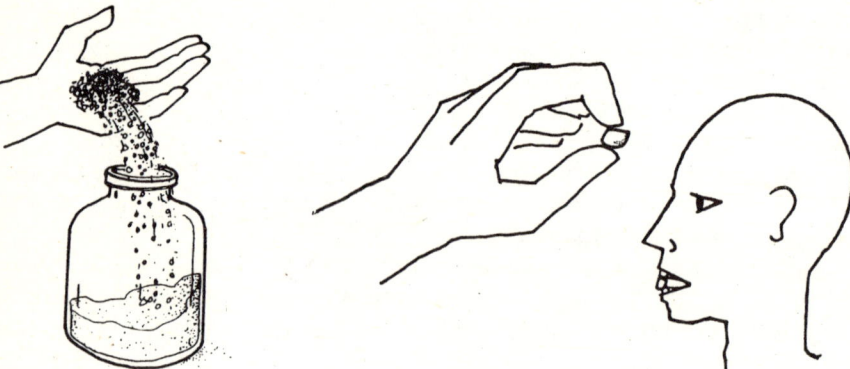

Figure 2.1 *Salt test*   **Figure 2.2** *Biting grain to determine storage suitability*

Other methods of determining fitness of grain for storage. The experienced person, without access to moisture meters, is able to obtain some idea of the suitability of grain for storage by the sight, feel and hardness of the individual kernels. Although this method is unreliable for determining moisture content accurately, it can prove useful in helping to decide whether grain may be stored with relative safety.

2.5 Sampling for loss assessment

2.5.1 **Background.** It is difficult to learn of the conditions in a total situation by direct measurement. This is because of the sheer size of the task involved and because the very process of taking a measurement probably changes what is being studied and measured, so that it is no longer true of the real situation. It is a more common practice to carry out tests on samples. If the results of these sample tests, however, are to be applicable to the situation as a whole, the sample itself must be truly representative.

For example, if information is required about the moisture content of a consignment of 100 tonnes of maize (the total situation), of uniform quality and uniform moisture content (a condition not possible in practice), the moisture content of any 1-g sample from any portion of the 100 million g in the 100 tonnes would be the same. The only errors which might arise would be from human factors or from defects in the measuring instruments. Sampling would present no problem; all samples, however selected, would contain maize of the same quality and the same moisture content.

In practice, however, any batch of 100 tonnes of maize would, initially, have variations of quality and of moisture content throughout its bulk, and even between individual grains. The variations themselves will change with time; insects and moulds will attack different parts of the maize. Moreover, heating, which produces "hot spots" and consequently more rapid deterioration in quality and changes in moisture content, will occur in local pockets.

Assessment of grain loss depends on being able to make accurate measurements on a representative sample. However accurately the characteristics of a sample in the laboratory may be determined, the results will be of little value if the sample was not representative of the original material. It is also true, of course, that however well a sample may represent the original, the final result will only be as valid as the accuracy of the instruments used and the competence of the operators. In practice, an acceptable degree of accuracy obtained at reasonable cost must be the goal.

2.5.2 **Sampling.** Taking samples that truly represent the original material is not easy. Samples may contain errors or bias. Thus, if the best-looking stack of grain in the field, the one nearest the home or the one that the farmer selects is chosen, bias will be present. The same will be true if samples are always taken near the granary entrance or where the grain looks good or bad. On the other hand, efforts to avoid bias may result in overcorrection. In avoiding bags within easy reach, those that are most difficult to reach may always be chosen. The solution is to remove the choice from the individual, and rely upon a table of random numbers. This is the basis of probability sampling, and the samples so obtained are termed probability samples. The advantages of a probability sampling plan are as follows:
(a) the degree of error due to sampling can be predicted;
(b) the number of samples required to obtain the desired accuracy in sampling can also be predicted; and
(c) the sample so obtained is certain to be representative of the original.

2.5.3 **Unit of observation.** The observational unit is the container, location or process from which a sample is taken and used to determine the loss. It is the smallest division or unit in which grain is held. It may consist of stacks in a field, small silos or granaries on a farm, or woven baskets — but a single basket rather than all a farmer's storage baskets; or individual bags rather

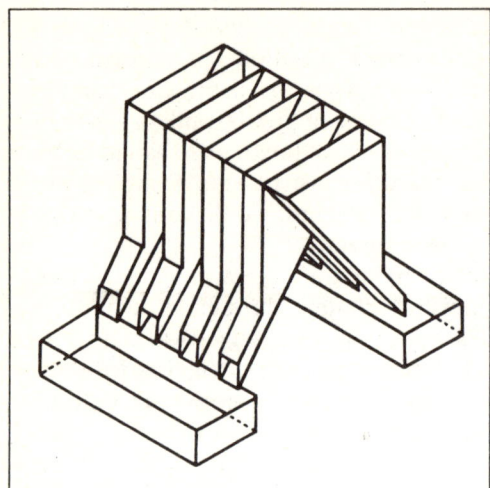

Figure 2.3 *Box divider for sampling*

than the whole warehouse. The value of the entire survey will depend on the accuracy with which the loss is determined for each observation unit.

To make sampling easier, the observational unit should be as small as possible. This makes it feasible to mix all the grain thoroughly and produce a representative sample by coning and quartering or by using a sample divider. Such a procedure may apply where the grain is in baskets or in stacks in the field. In silos or granaries, however, it may not be possible; and unless the operation is done with skill, the sample may contain a systematic error which cannot be removed by later calculation or analysis.

When a container is sampled as a unit, the assumption is that the defect, contamination or other characteristic to be identified is uniformly (or at least randomly) distributed within the unit. In practice this is usually not the case. For example, insects or mites, mouldy kernels, rodent depredation and insect-damaged kernels are more usually found in pockets or layers within the bulk.

With limitations of time and money and often with traditional cultural considerations, the best course is to design a method of sampling so that the grain will be as representative as possible of the undamaged material and the defects throughout the layers or pockets.

In any study the investigator must report what was done and why, so that the significance of the data can be understood by those who will use them.

2.5.4 **Taking samples.** Most quotations of loss assessment are based on a weigh-in/weigh-out system. For example, in a batch process all observational units are weighed after initial samples have been taken, and all observational units are reweighed after completion of the batch process and before final samples are taken.

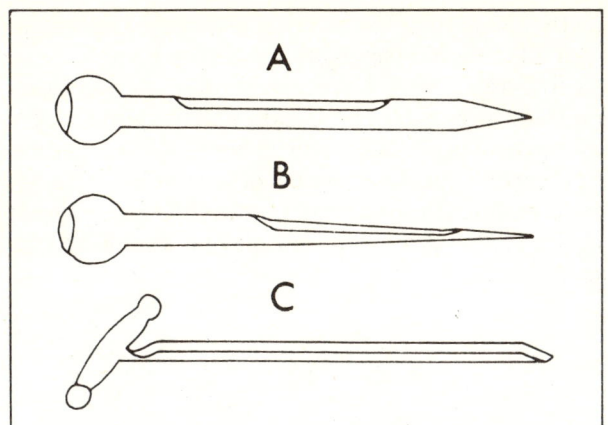

Figure 2.4 *Sack sampling spears*

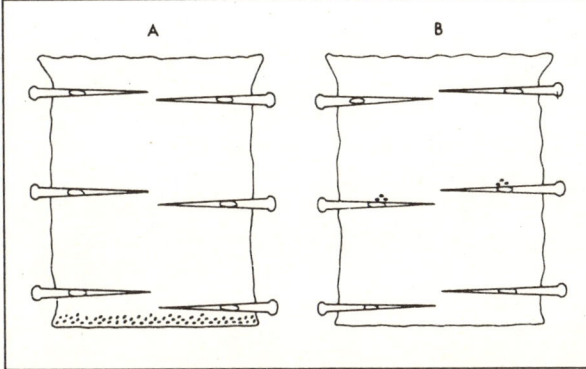

Figure 2.5 *Inadequacies of sack sampling spears (where black dots represent individual insects).*
A: large populations can be underestimated;
B: small populations can be overemphasized

In many cases, however, observational units must be selected, using random numbers. The general procedures given in the previous section on loss assessment would then apply.

2.6 Farm- and village-level grain flow pattern

The flow of grain from the field to the eventual consumer is often complex; it may be compared with the flow of water through a system of pipes. Losses of water can occur anywhere in such a system. It is important to establish the magnitude of individual water leaks so that the largest can be reduced.

Similarly, grain flows from producer to consumer and is subject to leaks or losses on the way. It is important to obtain a relative perspective in order to visualize the importance of the total grain lost in flowing from producer to consumer, compare this with the quantity lost at a particular point, and measure this loss as a proportion of the grain passing that point.

In any programme of loss assessment it is necessary to obtain as much local information on grain flows as possible: how and when the grain moves from harvest to consumer; the routes it follows and its holding patterns; and where and how processing is carried out. Each district or community area has a marketing system for food grains. It is essential to establish the flow lines and the quantities they carry, so that priority points may be established for the observation and measurement of losses.

3. Storage pests

3.1 Post-harvest microbiology

3.1.1 **Fungi.** The simplest structure of a fungus consists of a thread (or *hypha*) which grows inside the host material. Several hyphae produce a mat known as a *mycelium*. The asexual reproductive structures known as *sporangiophores* grow out of this mycelium and extend beyond the surface of the substrate, or host material. At the end of these sporangiophores is the sac (or *sporangium*) containing the individual spores. Sexual reproductive structures are less frequently observed. The classes of fungi most important in crop storage are the moulds or microscopic fungi, whose optimum temperature for development is above 20°C.

In order to multiply, fungi require water, oxygen and a suitable temperature. They also require food materials from the substrate; these are dissolved before absorption into the mycelium. Unless precautions are taken, stored crop products form an ideal substrate for fungus growth.

From an ecological point of view fungi may be divided into field and storage fungi.

Field fungi, such as *Alternaria, Fusarium, Cladosporium* and *Helminthosporium*, invade the seeds before harvest. These fungi develop only on seeds with a high moisture content (22-25 percent) and will die out under correct storage conditions.

Storage fungi, mainly *Aspergillus* and *Penicillium* species, develop on seeds with 12- to 18-percent moisture content.

3.1.2. **Some important post-harvest fungi.** There are many varieties of post-harvest fungi. Some of the most important include the following:

Aspergillus flavus. Grows on, and causes deterioration of, proteins, starches and oils; in particular, it reduces the quality of oil. Some strains produce the poisonous toxin aflatoxin, particularly in inadequately dried oil-seeds and cereals.
A. niger. Similar to *A. flavus* but the toxin produced is not so dangerous. The spore heads are black.
A. glaucus group. A very common group of moulds able to grow on substrates with very low moisture and high sugar content; usually the primary invaders of stored crop products.

Penicillium spp. Commonly associated with fruit rots. The mycelium is bluish-green and may be aerial or embedded in the substrate; widely dispersed.

Botryodiplodia spp. Attack fruits or seeds in the field and deterioration continues in storage. The mycelium is black in *B. theobromae*; spores are produced in enclosed pycnidia (cavities) at the surface of the substrate.

Fusarium spp. A widely occurring species, as a fungus associated with storage rots, and as a pathogen causing blights and blasts of cereals and sugar cane. May survive in the seed and may continue growing during storage. Some species produce toxins on stored maize that has not been dried to a safe moisture content. Two species belonging to the *Phycomycetes* are also common on stored products.

Rhizopus otrhijus. Very widespread and reproduces sexually in characteristic sporangia on many crops, but is not a primary invader.

Mucor pusillus. A fungus associated with spoilage and decay. Strongly thermophilic; for example, it can survive the high temperatures of fermenting cocoa.

3.1.3 **Controlling fungal growth on stored products.** The primary cause of fungal deterioration of stored products is the presence of excessive water. This leads to relative humidities exceeding 70 percent in the air within the bulk of the stored product. Many fungal mycelia will develop at this relative humidity, leading to an increase in biological activity and increased deterioration.

The 70-percent level of relative humidity (rh) (see Section 6.3) has come to be regarded as a "safe" level; the moisture content of commodities in equilibrium with this rh indicates the upper limit of moisture content for safe storage (see Table 1).

TABLE 1. Values of moisture content equilibrium at 70 percent relative humidity and 27°C[1]

Commodity	mcwb	Commodity	mcwb
Maize	13.5	Groundnuts (shelled)	7.0
Wheat	13.5	Cottonseed	10.0
Sorghum	16.0	Cocoa beans	7.0
Millet	16.0	Copra	5.8
Paddy	14.0	Palm kernels	5.7
Cowpeas	13.5	Gari (yellow)	13.6
Beans	15.0	Gari (white)	12.7

[1] Determined from prolonged exposure to controlled atmosphere — conditions which do not always apply to stored products.

3.2 Pest biology and identification

3.2.1 **Ecology.** Many species of insects are associated with stored produce. Of these, only some directly damage the produce. The produce can support an insect community including scavengers, predators and parasites as well as the primary pests. Each species will show different behaviour, tolerances and preferences with respect to:
- commodity
- moisture and temperature
- state of the produce (e.g. intact, damaged or milled)

For a particular locality, commodity and storage method there are usually only a few species which are important pests.

3.2.2 **Identification.** It is important to identify the main pest species in order to:
(a) assess whether the insects found are likely to cause serious damage and to merit control; and
(b) choose an appropriate control technique, since many treatments are selective in their action.

3.2.3 **Biology.** The effectiveness of control measures may be greatly increased by applying an elementary knowledge of the biology of pest species. For example, what is the likely source of infestation? Does the pest have a resistant phase? It is mobile enough to reinfest? What are its tolerances?

3.2.4. **Reference collections.** The majority of storage pests are small and difficult for non-specialists to identify. The accompanying notes are only an introduction.

Field-workers can identify insects most easily by comparing them with a prepared reference collection of the main pests to be found in their particular area.

The most common pests should be collected and sent for specialist identification. Larvae, caterpillars and pupae should be preserved in 70-percent ethyl alcohol. Adults should be submitted as found. Specimens should be labelled with the locality in which they were collected, the date and the produce concerned.

3.3 Descriptive notes

In the descriptions that follow the title of each pest in sections 3.3.1-3, the lettered notes correspond to the following system:
(a) recognition; (b) commodities attacked; (c) pest status.

3.3.1 Major primary pests
Of cereals

1. Maize weevils/rice weevils: *Sitophilus* spp. (Col., Curculionidae).

(a) distinguishable from all other common storage pests by the long beak (or rostrum) characteristic of all the weevils; 2.5 to 4 mm long, dark brown, sometimes with four lighter spots on the wing cases;
(b) maize, rice, sorghum, wheat;
(c) the most important primary pests of cereals in the humid tropics; attacks undamaged grain; often infests before harvest. The larvae develop within the grain, leaving a characteristic round hole on emergence.

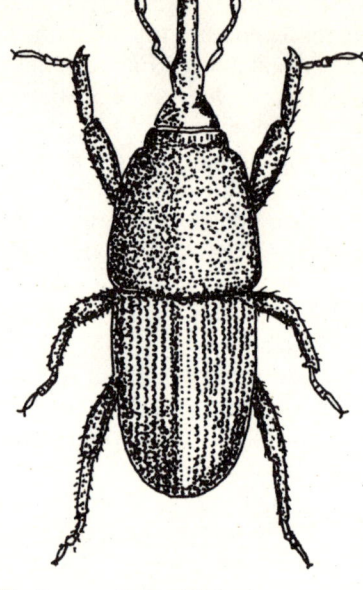

Figure 3.1 Sitophilus *spp.*

2. Angoumois grain moth: *Sitotroga cerealella* (Lep., Gelechiidae).

(a) a small cream- or fawn-coloured moth, sometimes with a small black spot on the forewing; the wings are very narrow and fringed with long bristles; the sharply pointed tip of the hindwing is characteristic;
(b) sorghum, maize, wheat, rice;
(c) Sitotroga *replaces* Sitophilus *as the main pest in the more arid areas. Damage may be very serious in maize stored on the cob; damage is more limited with shelled grain, as the moths do not penetrate more than a few centimetres from the surface. The developing larvae cause all damage, as the adults do not feed.*

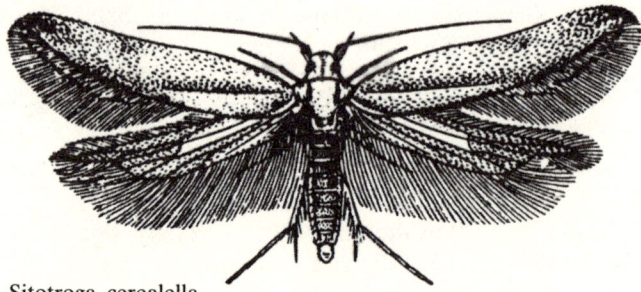

Figure 3.2 Sitotroga cerealella

3. Lesser grain borer: *Rhizopertha dominica* (Col., Bostrychidae).

(a) a small, almost cylindrical beetle, with the head "tucked" under the thorax so that it is invisible from above; the thorax has a prominent pattern of tubercles, as shown. Dinoderus spp. (also Bostrychidae) are also sometimes found in maize; they have a similar cylindrical shape but are markedly shorter and stouter;
(b) sorghum, maize and other cereals; cassava;
(c) a major primary pest in drier tropical regions. The Bostrychidae are adapted to boring into hard substances such as wood and are capable of attacking previously undamaged grain where they can cause serious damage. Bostrychidae may also sometimes be found attacking the timber of storage structures.

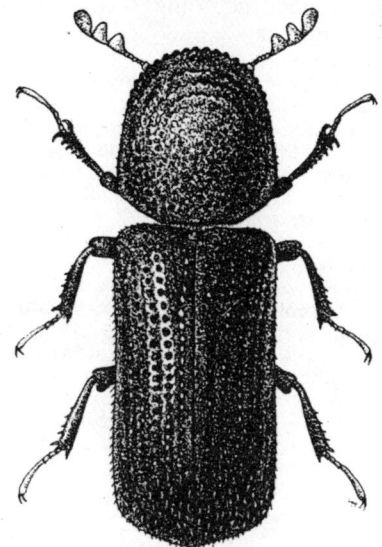

Figure 3.3 Rhizopertha dominica

4. Larger grain borer: *Prostephanus truncatus* (Col., Bostrychidae).

(a) very similar but slightly larger than the lesser grain borer. Larvae and adults cause damage to various commodities;
(b) maize and other cereals, cassava and groundnuts;
(c) this pest was accidentally introduced from South or Central America into Africa, where it causes heavy damage, especially to maize stored on the cob and dried cassava. Weight losses caused by this pest are 3 to 5 times higher than those caused by the normally occurring pest. Countries infested at present are the United Republic of Tanzania, Kenya and Togo. It is expected that this pest will spread to other countries.

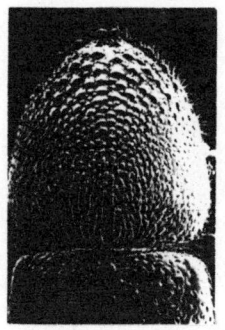

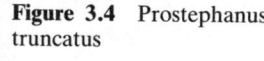

Figure 3.4 Prostephanus truncatus

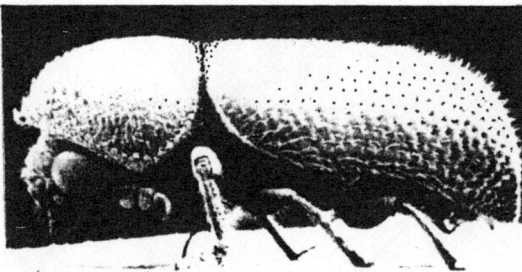

Of grain legumes
Bruchids (including the bean weevil): *Callosobruchus maculatus* (Col., Bruchidae).

(a) rather stout beetles; active, with long legs and antennae; antennae not "clubbed"; wing areas often mottled, spotted or otherwise marked; last segment of abdomen just visible beyond wing cases. The various genera and species in this group are difficult for non-specialists to identify;
(b) preferred commodity varies with species. Callosobruchus species feed on cowpeas, pigeon peas and green gram; Acanthoscelides *on* Phaseolus *beans;* Caryedon serratus *on groundnuts;*
(c) several species are major pests on their characteristic crops, especially Callosobruchus maculatus *on cowpeas. Infestation often commences in the field before harvest; larvae develop hidden within the bean.*

3.3.2 Locally important pests
Pyralid moths: *Ephestia* spp., *Plodia interpunctella*, *Corcyra cephalonica* (Lep., Pyralidae).

(a) all have the general outline shown, with broader wings than Sitotroga *and a shorter fringe of bristles.* Ephestia *spp. have dark forewings, sometimes indistinctly banded, and paler hindwings;* Corcyra *is uniformly dark grey-brown;* Plodia *has forewings cream at the base and red-brown on the outer half;*
(b) various species on cereals, milled cereal products, groundnuts, dried fruit;
(c) can be major primary pests, and important on flour and other products; larvae, which are free-living caterpillars, spin silk as they move (which is both a problem in itself and often the first visible sign of infestation).

Figure 3.5 Callosobruchus maculatus

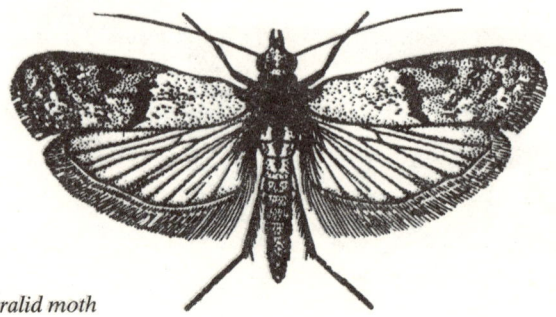

Figure 3.6 *Pyralid moth*

Flour beetles: *Tribolium, Gnatocerus, Palorus,* etc. (Col., Tenebrionidae).

(a) elongated, reddish-brown beetles, active; Gnatocerus *spp. are recognizable from the small, upward pointing horns on the head of the male;*
(b) cereals (especially after damage by primary pests), groundnuts, milled cereal products;
(c) can attack intact grain via the embryo, but infestation is usually more serious on damaged or milled products, where they can be major pests.

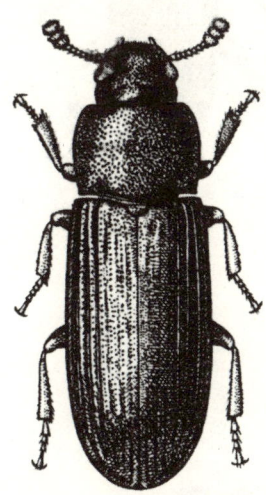

Figure 3.7 Tribolium

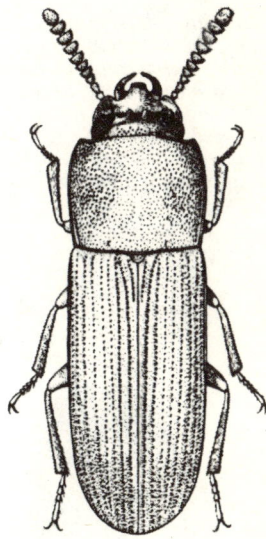

Figure 3.8 Gnatocerus

Khapra beetle: *Trogoderma granaria* (Col., Dermestidae).

(a) dark brown, "marbled" with lighter bands; very finely hairy; 3 mm long; larvae with long conspicuous bristles. Easily confused with other dermestids; if an infestation is suspected, identification by an expert should be sought;
(b) groundnuts, cereals, grain legumes;
(c) major pest in drier areas; important partly because larvae can enter a resistant resting phase (lasting up to several years); difficult to eradicate.

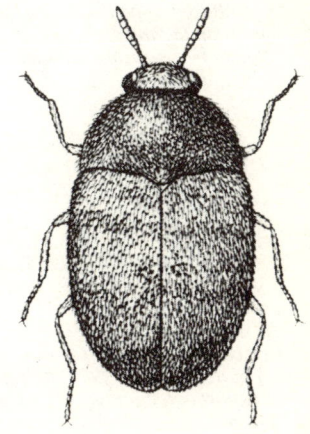

Figure 3.9 Trogoderma granaria

Merchant and saw-tooth grain beetles: *Oryzaephilus* spp. (Col., Silvanidae).

(a) active, dark-brown beetle, about 4 mm long, markedly elongated and flattened in shape; recognizable by the six prominent teeth on each edge of the thorax;
(b) cereals (especially rice), cereal products, oil-seeds;
(c) secondary pests; can be particularly important on milled products.

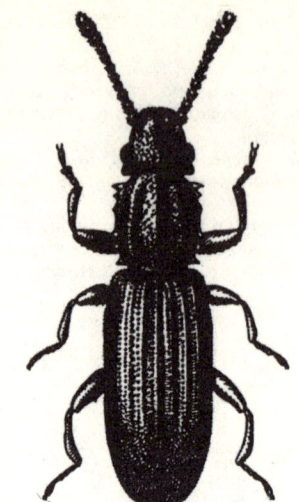

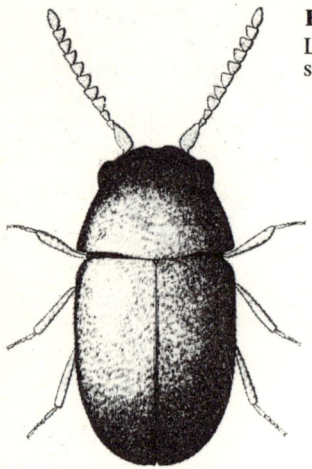

Figure 3.11 Lasioderma serricorne

Figure 3.10 Oryzaephilus *spp.*

Tobacco beetle: *Lasioderma serricorne* (Col., Anobiidae).

(a) reddish-brown, finely hairy, 2 to 4 mm long; head curved under thorax;
(b) tobacco and cocoa; secondary on cereals and grain legumes;
(c) can be important on any of the above commodities.

Coffee-bean weevil: *Araeocerus fasciculatus* (Col., Anthribidae).

(a) grey-brown, usually slightly mottled; similar in shape to a bruchid, with the tip of the abdomen exposed; distinguished from bruchids by the loose, three-segmented club on the antennae;
(b) coffee, cocoa; secondary on cereals;
(c) not usually very damaging but its presence in export consignments can lead to their rejection.

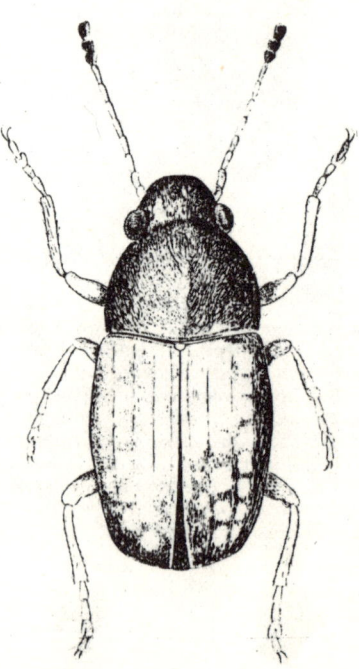

Figure 3.12 Araeocerus fasciculatus

Dermestids (skin beetles): *Dermestes* spp. (Col., Dermestidae).

(a) larger (5- to 10-mm) beetles, usually black or black and white; larvae bristly as in Trogoderma;
(b) animal products;
(c) major pests of dried fish, dried meat, hides.

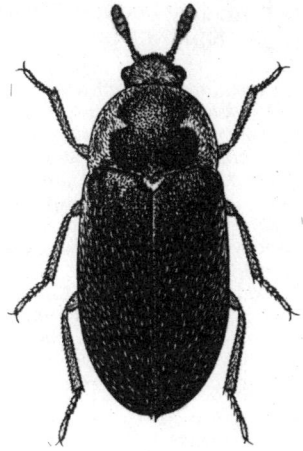

Figure 3.13 Dermestes *spp.*

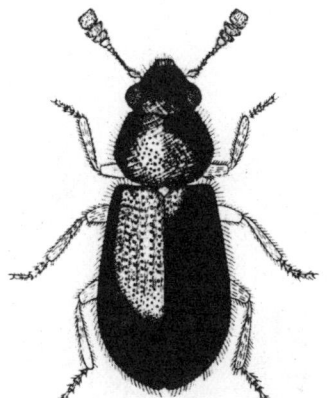

Copra beetle: *Necrobia rufipes* (Col., Cleridae).

(a) 4- to 7-mm, metallic green body, reddish legs;
(b) copra, palm kernels, animal products;
(c) serious pest only on mouldy produce; can be common on dried fish, etc.

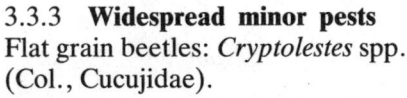

Figure 3.14 Necrobia rufipes

3.3.3 Widespread minor pests
Flat grain beetles: *Cryptolestes* spp. (Col., Cucujidae).

(a) very small (1 to 2 mm), flattened, red-brown in colour;
(b) cereals, cereal products, cowpeas, cocoa;
(c) can become very abundant, especially in flour or damaged grain.

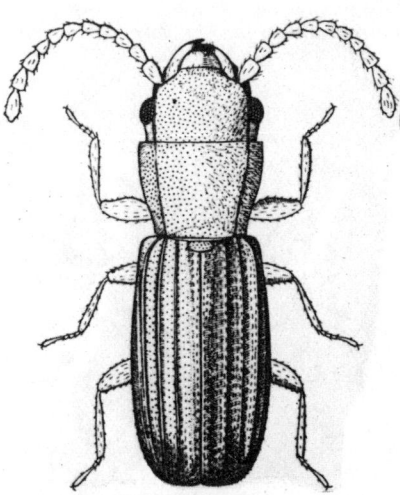

Figure 3.15 Cryptolestes *spp.*

Sap-feeding beetles: *Carpophilus* spp. (Col., Nitidulidae).

(a) small, active beetles which may be brown or black in colour; sometimes with orange-brown patches on the wing cases; distinguishable from other storage pests by the last two segments of the abdomen, which are not covered by the wing cases and are clearly visible from above. In the related Brachypeplus *species, three abdominal segments are visible;*
(b) damp grain, palm kernels, copra, cocoa;
(c) very widespread and sometimes abundant; common at harvest in cereals; only damaging on stored grain where produce is already damaged or mouldy.

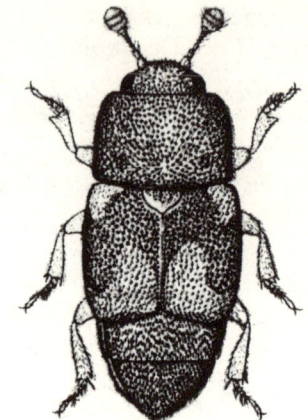

Figure 3.16 Carpophilus *spp.*

3.3.4 Other arthropods
Ants (Hymenoptera, Formicidae) and termites (Isoptera).

Can be abundant in farm stores; usually act only as scavengers and so rarely need controlling; termites may cause severe damage to timber structures.
   Ants can be controlled with insecticidal dusts applied on the (usually distinct) trails to the communal nests; timber may be protected from termites (and fungal rots) by regular application of old engine oil.

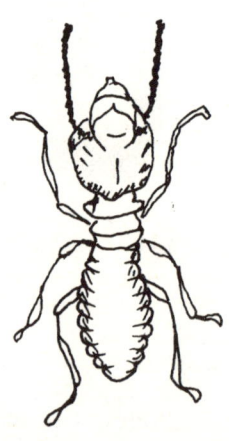

Figure 3.18 *Termite*

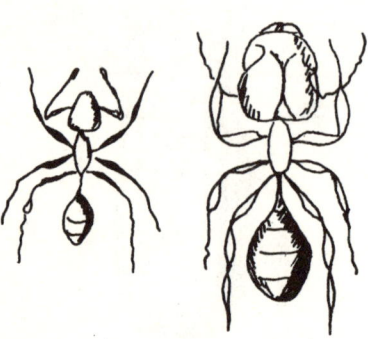

Figure 3.17 *Ants*

Parasitic wasps (Hymenoptera).

Very tiny insects (most less than 2 mm long), usually with four clear wings. They are beneficial, parasitizing the eggs and larvae of various moth and beetle pests. Can help to reduce pest increase in some situations.

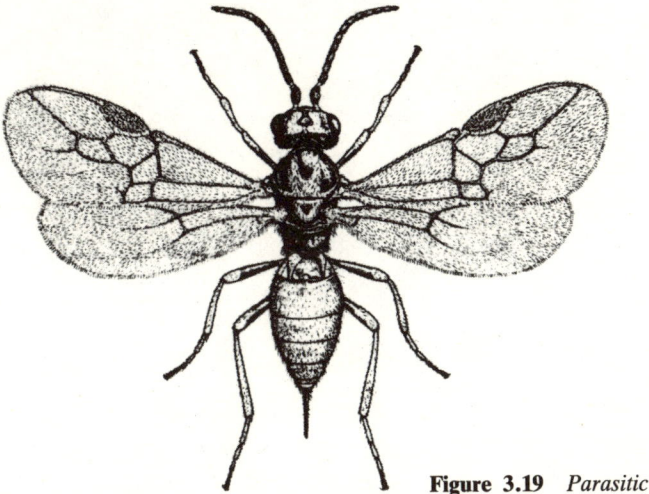

Figure 3.19 *Parasitic wasp*

Mites.

Mites belong to the class of Arachnida (subclass Acarina) and may be distinguished from insects by the possession of eight legs and an apparently unsegmented body. Those found on stored products are extremely small, 0.2 to 1 mm in length, and are easily overlooked.

Some species are predators on the eggs of moths or on other mites, but many are serious pests of flour and other processed foods. The types are difficult to distinguish, but the pest species are smaller than the predatory ones (a hand lens is usually required to see them), whitish in colour and slow-moving. Their importance as pests of stored products in the tropics has not been properly investigated, but if found in very large numbers they should be regarded as pests.

Phosphine fumigation kills mites but other insecticides may be less effective. If there is a problem with mites it is important to choose a chemical (e.g. an acaricide) that specifically states that it is effective against them and approved/recommended for use on stored products.

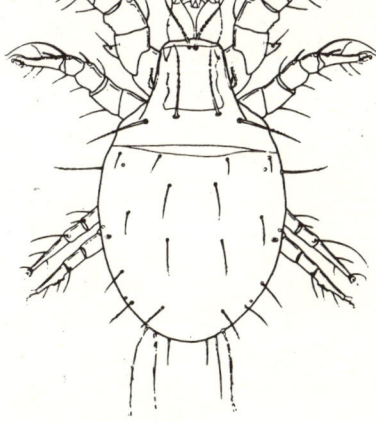

Figure 3.20 *Mite*

4. Loss assessment

4.1 Introduction

Food-loss assessments provide the basis of programmes for reducing post-harvest losses. Assessments may be made by *surveys* of both traditional methods and improved methods, and be followed by quantitative, technical and financial comparisons. *Trials* may be conducted to determine the acceptability of improved storage structures or methods of operating. A distinction should be made between loss surveys and field studies or trials; but both may seek to compare traditional and improved methods for reduction of losses.

4.2 Definitions

Many of the terms used in loss assessment measurements are given in Section 1.1.

4.3 Surveys

Three types of loss surveys can be identified.

4.3.1 **General survey.** A preliminary examination of specific problem points, and on-site appraisal of measures likely to reduce losses. This type of survey should be conducted before any loss-reduction project commences. As a result, the post-production system will be fully understood, the point at which significant losses may arise identified, and the causes of these losses suggested. (In addition, all relevant available data from other sources on losses should be collected and collated. If possible, the expected losses should be roughly estimated.)

4.3.2 **Pilot survey (non-randomized survey).** In this type of survey, a quantitative approach is based on established sampling techniques, but a completely scientific sampling design is not followed.
    The sampling techniques adopted will determine the reliability of the survey results (see Section 2.5). A completely randomized survey is very costly, although difficulties arising from lack of cooperation by farmers and inaccessibility of sites are usually the decisive factors in choosing villages and

farms for the sample. A statistically valid sampling of stored unshelled produce is rarely possible unless the storage container is emptied. This is often unacceptable to the farmer; it also disturbs the pest populations and stratifications that exist in the container. This storage container and its contents cannot then be used later for loss assessment measurements. Therefore, in pilot surveys on storage losses, random samples are usually taken from the produce which *can* be reached.

In spite of these limitations, pilot surveys on storage, even when they may not be statistically sound, can provide valuable data on the losses occurring and allow their progression over time to be monitored.

The aim of a pilot survey is to establish an estimate of the losses and provide data on their causes. Improvements may then be introduced, which need to be monitored and adjusted as more information becomes available.

4.3.3 **Reliable survey (randomized survey).** In this type of survey the objective is to obtain statistically reliable quantitative data on losses at village, regional or national level. A stratified random sampling programme and statistically acceptable method of sample analysis are essential. Such procedures are costly and require a large number of personnel specially trained for the job.

Such surveys are best suited to the processes of harvesting, threshing, drying and processing. They are less suitable for evaluating losses during storage because of problems arising from the processes of biological deterioration.

4.4 Field trials

Field trials are frequently used to compare the losses that occur with traditional and improved practices.

Three types of trials may be conducted.

4.4.1 **Equipment testing.** Newly developed or purchased equipment must be tested for its suitability to harvest, thresh, dry or process the locally produced crops. Once a certain type of equipment has been found to be effective, it is essential to determine how it performs when used by farmers.

4.4.2 **Storage simulation trials at research stations.** Great caution must be exercised when interpreting results from this type of trial because the situation at research stations is very different from that on the farms.

4.4.3 **Trials at farm level.** These trials are conducted to meet two objectives.

Alternative post-production practices can be evaluated for their effect on the level of losses. The trials are conducted with the cooperation of farmers in their own fields and villages or, if this proves impossible, at local research stations or experimental farms; these results, however, should be treated with due caution.

Potential improvements are introduced and evaluated when used by a group of farmers, without previously being fully tested by research stations. This type of trial is extremely useful in evaluating methods which have already proved successful in other regions or countries. During the trial, adjustments may be made to improve the performance of the equipment. Suitable practical training programmes are arranged for the farming community.

4.5 Validity of survey and trials work

The result of surveys and evaluations is only valid for the conditions under which they were conducted. There is almost always a seasonal effect, which can only be determined over a number of years, on the level of losses.

A more practical solution to this problem is to conduct a survey to identify where in the system the losses are heavy and the magnitude of these losses. Improvements are introduced and are intensively monitored and evaluated. The improved and unimproved methods are compared, from the viewpoint of reducing losses, in the same season and same locality.

Such long-term surveys, although providing reliable information, are very costly and do not, of course, lead directly to a loss reduction.

4.6 Loss assessment in cereals and pulses

4.6.1 **Loss during harvesting, in-field drying, stacking, transport, threshing, drying and cleaning.** During these activities, all losses caused by biological agents should be adjusted to a dry-weight basis. Other losses are usually expressed in terms of weight of material at 14-percent moisture content.

Both these methods of presentation require calculations that may lead to confusion in interpreting the results; but if the losses are expressed as percentages (rather than adjusted weights), and the basis of presentation is stated, the results can be compared with other loss-assessment figures.

Losses in harvesting, in-field drying and stacking operations are expressed as a percentage of yield. The yield is defined as the *obtained yield*, which is the maximum quantity of clean grain minus the losses being assessed. During threshing and cleaning, losses are expressed as a percentage of the grain input to the operation.

Farmers' practices in time of harvesting, duration of in-field drying, and stacking have a marked effect on the level of losses. Harvesting losses generally increase with any delay in the harvest beyond the time considered best by the farmer.

Losses during in-field drying, transport, sun-drying and cleaning tend to be low, while threshing losses depend very much on the methods used and the time of harvest.

4.6.2 **Loss during storage at farm and village level due to insects and moulds.** The loss in weight during storage must always be related to the quantity in store at the time of assessment. There are three methods of assessing losses during storage, and others are being developed.

In the *standard volume/weight method*, the dry weight of a standard volume of grain is measured by a standard method at the beginning of the storage period and is compared with the dry weight of the same volume of grain after a certain storage period. The principle underlying the method is that the main storage pests develop inside the grains. The shape of the grains will remain intact and the damaged grains will occupy the same volume but will weigh less. There are also, however, surface-feeding insects. Their feeding may affect the volume of the grains, so that more grains may fit in the same volume, leading to unreliable results.

The dry weight of a standard volume of grain depends on moisture content and variety. A standard base-line curve for the dry weight of a fixed volume of grain as moisture content changes must be prepared at the beginning of the storage period for all varieties for which losses are being assessed. This is a major problem of the method. The method has been used successfully without base-line data in very dry climates where moisture content changes are small during the storage period.

In the *count and weight method*, the damaged and undamaged grains in a sample of 100-1 000 grains are counted and weighed. The weight of the sample is compared with the weight it would have registered in the absence of damage. The basic equation is:

$$\% \text{ weight loss} = \frac{UaN - (U + D) \times 100}{UaN}$$

where U = weight of undamaged fraction in sample
N = total number of grains in sample
Ua = average weight of one undamaged kernel
D = weight of damaged fraction in sample

The percentage weight loss must be adjusted to 14-percent mcwb, or the moisture content should be stated.

The percentage weight loss can also be calculated as:

$$\text{percentage damaged grains} \times \frac{\text{(mean weight loss per grain)}}{\text{(mean weight undamaged grain)}}$$

$$= p \times \frac{MWL}{MW}$$

Let Nu = number of undamaged grains
Nd = number of damaged grains
U + D = as above, then

$$P = \frac{Nd}{Nu + Nd} \times 100$$

and

$$\frac{MWL}{MW} = \frac{\frac{U}{Nu} - \frac{D}{Nd}}{\frac{U}{Nu}}$$

$$= 1 - \frac{D\ Nu}{U\ Nd}$$

$$= \frac{U\ Nd - D\ Nu}{U\ Nd}$$

Percentage weight loss $= P \times \dfrac{MWL}{MW} = \dfrac{(Nd \times 100)}{(Nu + Nd)} \times \dfrac{(U\ Nd - D\ Nu)}{U\ Nd}$

$$= \frac{(U\ Nd - D\ Nu) \times 100}{U\ (Nd + Nu)}$$

This formula does not require the value of the mean weight of undamaged grain.

The main disadvantages of this method are that:

(a) insects may show a preference for grains of certain dimensions, composition or moisture content, and consequently the mean weight of insect-damaged grains before damage may be different from the mean weight of grains in the undamaged sample; and

(b) a visibly undamaged grain may have a hidden infestation which leads to an underestimate of the loss.

Because of these possible sources of error, negative weight-loss figures may be obtained at low infestation levels.

In the *converted percentage damage method,* the percentage damage is converted into weight loss by multiplying by a factor which is constant for a particular commodity. The method is of limited reliability and is only used when the other two methods cannot be used.

5. Pest control

5.1 General

Before attempting to apply control measures it is essential to identify the pest concerned and to understand how it is a threat to the safe storage of the commodity.

Preventing infestation is always preferable to controlling infestation that has assumed serious proportions. The potential sources of infestation must be known so that the build-up of pests during storage can be controlled more easily and economically.

The type of storage structure influences the susceptibility of its produce to pest development. It also affects the method of control that will be most economical. A brief description of storage structures is given in Section 5.6.

Figure 5.1 *Potential sources of infestation*

5.2 Losses caused by insects

5.2.1 **Weight loss.** As they develop on a commodity, insect pests feed continuously. Estimates of the resultant loss vary widely with the commodity, the locality and the storage practices involved. For grains or grain legumes in the tropics, stored under traditional conditions, a loss in the range of 10-30 percent might be expected over a full storage season.

5.2.2 **Loss in quality/market value.** Infested produce is contaminated with insect debris and has an increased dust content. Grains are holed and often discoloured. Food prepared from infested produce may have an unpleasant odour or taste.

Prices in traditional markets are relatively insensitive to pest-damage. Centralized marketing and distribution of produce usually depend, however, on a grading system which penalizes infested produce.

Export crops such as coffee, cocoa and groundnuts call for particularly high quality standards.

5.2.3 **Promotion of mould development.** Insects, moulds, and the grains themselves produce water in *respiration*, i.e. a breakdown of carbohydrate substrate. In humid conditions, without adequate ventilation, mould development and "caking" can spread rapidly, causing severe losses.

5.2.4 **Reduced germination in seed material.** Damage to the embryo of the seed will usually prevent germination; some storage pests prefer to attack the embryo.

5.2.5 **Reduced nutritional value.** Removal of the embryo by storage pests will tend to reduce the protein content of the grain.

5.3 Sources of infestation

5.3.1 **Survival from last season.** Insect pests are able to survive from one season to the next in a variety of situations, as follows:
(a) *infested residues* from the previous year, stored at home or on the farm;
(b) *the structure of the store itself,* in:
    (i) the thatch, bamboo or timber of a traditional crib,
    (ii) cracks in the wall of a silo, rumbu or warehouse, or
    (iii) old sacks and dunnage; [1] and
(c) *natural habitats,* such as:
    (i) under the bark of trees,
    (ii) in rotting wood, or
    (iii) in seed pods.

---

[1] Some pests are able to survive in a resting state without food, or, as with *Trogoderma granaria*, by feeding on minimal accumulations of dust, flour and debris.

Figure 5.2 *Survival of insect pests*

5.3.2 Infestation of fresh produce. Such produce can be infested by:
(a) active migration to the crop maturing in the field, from home, farm-store, warehouse or "bush"; or
(b) contamination when material is put into a store that is already infested.

When crops mature in the field they may be infested with storage pests:
(a) maize, sorghum and other cereals can be infested by maize and rice weevils (*Sitophilus* spp.);
(b) cowpea and other grain legumes can be infested by bean weevils (Bruchidae); the larvae of these pests, developing within the seeds, will be carried into the store with the produce and will continue to breed.

The severity of field damage can be considerably affected by the crop variety and cultural practices.

5.4 Pest build-up in store

In the humid tropics, conditions in stores may be highly favourable to the development of many species of storage pests. At 27-30°C and 70- to 90-percent relative humidity, on appropriate substrates, potential rates of increase are very high; examples include 25 times increase per month for the rice weevil (*Sitophilus oryzae*), 50 times per month for the bean weevil (*Callosobruchus maculatus*) and 70 times per month for the red flour beetle (*Tribolium castaneum*).

Competition, predation and parasitism may reduce the number of insect pests developing.

Dry conditions can significantly slow down the rates of development.

In general, a pest problem can be expected throughout the season in the more humid zones, but in semi-arid savannah areas pest activity usually subsides during the dry season.

5.5 Factors affecting choice of storage method and pest control measures

At every level of storage there will be a choice of different storage methods:
- domestic (subsistence)
- farm (cash cropping)
- community (commercial)
- centralized (national)

The suitability of a particular method will depend on a number of factors, including:
- scale of operations
- value of the commodity
- capital and running costs
- availability of materials and expertise
- climatic conditions
- pest problems

Following are summary descriptions of each type of structure, with emphasis on the techniques for pest control that are possible within each system.

The methods described should not be taken as recommendations; instead, they seek to suggest a way of evaluating the various options available in a particular situation.

5.6 Types of storage and implications for pest control

5.6.1 Traditional cribs.
These are characterized by lower levels of ventilation than in the improved designs; maize is often stored in the husk, sorghum and millet on the head.

The crop must remain longer in the field to dry sufficiently to avoid moulding. Losses in the field due to birds, rodents and "lodging" will be severe, especially in humid areas.

The crib itself may not be secure against rodents.

Husk (for maize) may provide considerable protection against insects for traditional varieties. If an insecticide is used its efficiency may be reduced by the presence of the husks, although there is conflicting evidence; insecticides used are usually dusts.

In a "closed" type of crib (e.g. basketwork), insecticide may be relatively persistent; but since such cribs have poor penetration there is no scope for reapplication.

The capital cost of the structure is low but its life may be short.

Figure 5.3 *Traditional cribs*

5.6.2 **Improved cribs.** These cribs are well ventilated, allowing harvest at high moisture content. (Early harvest also reduces field loss.)

Such a crib can protect against rodents.

Ventilation more or less eliminates the mould problem but there may be superficial germination in very wet conditions.

Husks must be removed, because of the high moisture content, exposing grain to insect attack; insecticide treatment with dust or spray is therefore necessary in most localities.

Insecticides admixed initially will tend to break down rapidly, but can be reapplied, at least to the outside of the crib. Penetration to the inside is thus improved.

The capital cost is low to moderate, depending on the materials chosen. The durability of the cribs will also depend on materials. The recurrent cost is the pesticide.

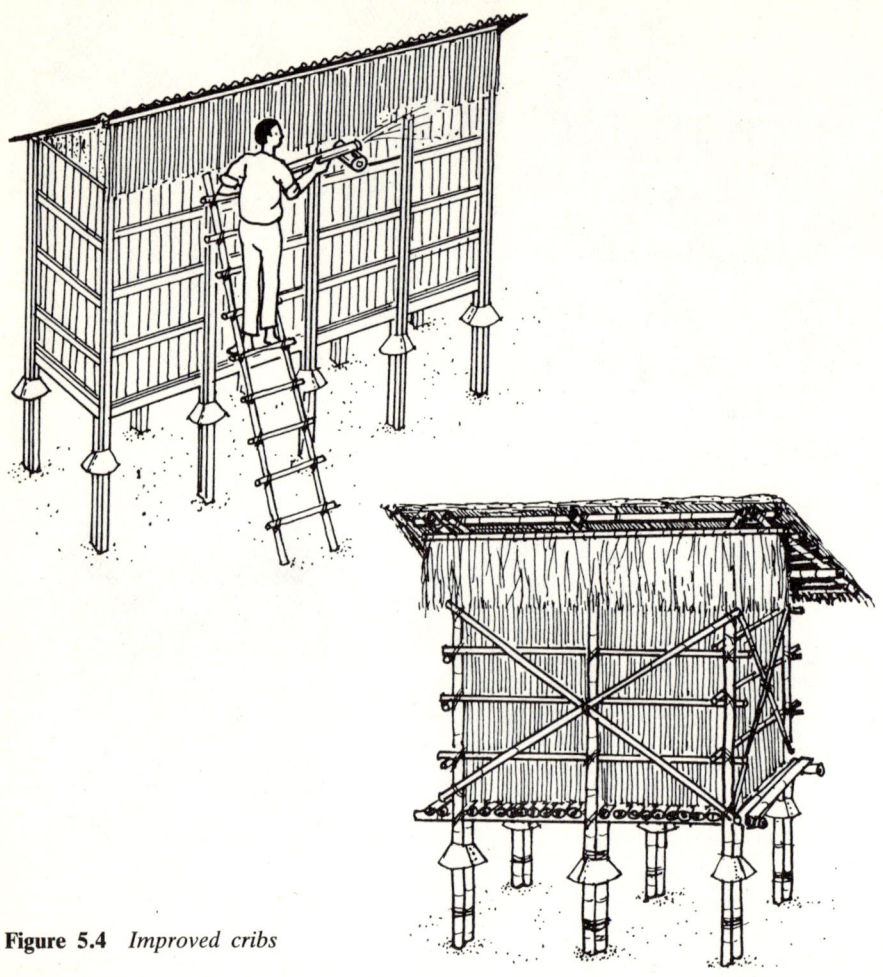

Figure 5.4 *Improved cribs*

5.6.3 **Silos.** Silos (traditional and improved) are unventilated structures storing bulk grain.

Produce must be very dry initially; artificial drying is obligatory in humid zones.

Rodent damage can be eliminated.

Moulding is very likely if condensation occurs; heating and cooling each day promote migration of moisture and local caking which can spread rapidly.

Silos require frequent inspection to guard against caking and may require artificial ventilation (which is not feasible at the rural level) or emptying and redrying.

Insect control in silos is potentially good. A suitable structure can be fumigated initially and then sealed against reinfestation; admixed insecticides (i.e. dusts) should prove comparatively persistent.

Low moisture content results in slow insect development.

With good management, silos are effective; but with poor management there is a risk of rapid and total loss of crop.

Bulk-handling equipment is required to install the larger types of silos.

The capital costs are high, and sometimes very high, depending on the materials used. The recurrent cost of drying may also be high, and considerable labour is required to collect fuel at harvest time.

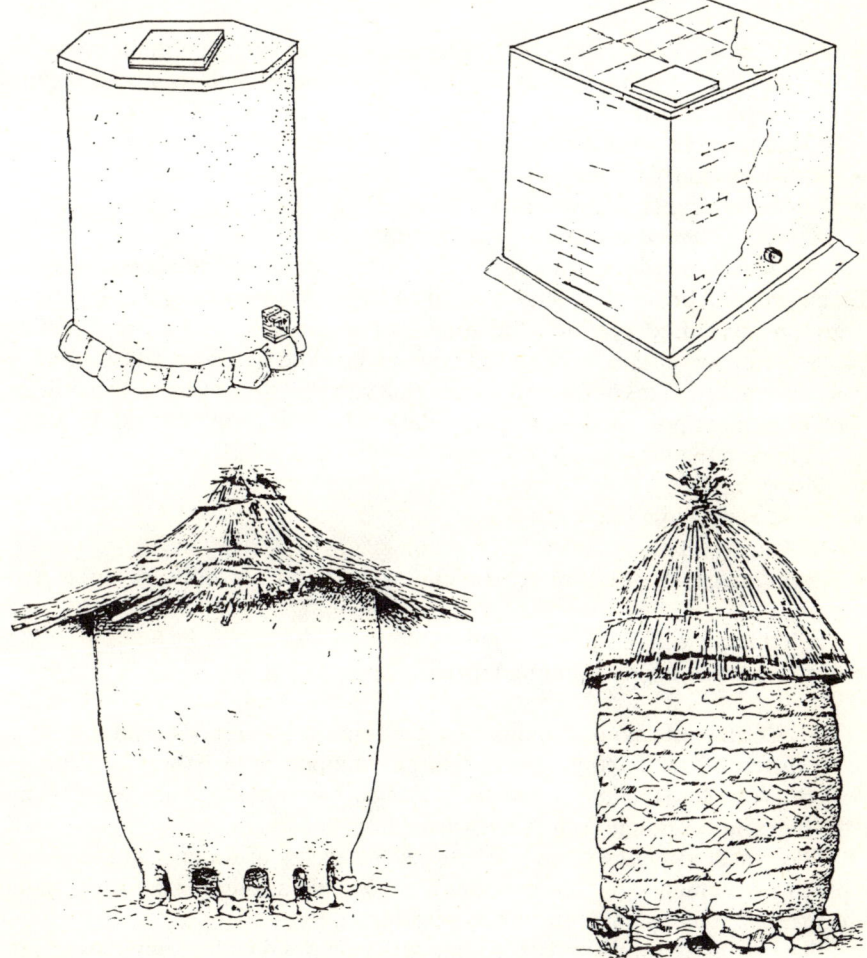

Figure 5.5 *Silos*

5.6.4 **Warehouses (and bag-stores in general).** These require some initial drying, but have higher tolerance than bulk storage.

Rodent control is possible and bagged produce can be fairly well protected against pests.

For valuable produce it may be justifiable to fumigate and prevent reinfestation. Spraying is far more effective than in ventilated structures; insecticides will be comparatively persistent.

Handling in warehouses does not necessarily require specialized equipment.

5.7   Losses caused by rodents

Rodents cause food loss by consuming grains and also by contaminating far more than they consume. They also spread diseases which may be transmitted to humans.

Three species of rodents are major pests of stored products:
- *Rattus rattus* (black rat) and *Rattus norvegicus* (brown rat)
- *Mus musculus* (house mouse)
- *Praomys natalensis* (multi-mammate rat)

Rats become active after dark or when the premises have become quiet. Black rats and brown rats have a habit of following set routes as they move between the stored product, the source of water and their normal resting place. After some time these routes become marked by greasy traces and are easily identified. This habit also means that rats will avoid unfamiliar objects such as traps or poisoned food, particularly when they are first laid down.

Clues to the presence of rats include:
- droppings
- loose earth from burrowings
- footprints on dusty floors
- greasy marks on set routes of travel, e.g., on beams or along electrical wiring
- holed sacks with grain escaping
- gnawing damage to building fabric

5.7.1 **Control methods.** Eliminating a single rat from a domestic house is very different from controlling a large number in a group of storage warehouses. Knowledge of the habits of rats is important in establishing effective and economic control measures.

The most effective control is to prevent access of rodents to the store. This can be achieved by "rat-proofing", preferably as the warehouse is being constructed, but also as a feature to be added later.

The main methods for controlling an established rodent population fall into the categories of *mechanical* and *chemical*.

The principal *mechanical method* of control is by trapping. The cage trap is preferable for godowns and should be set on the rat run. After being placed in position and left open for several days, unbaited and unset to overcome the

Figure 5.6 *Rat proofing storage warehouses*

rats' shyness of new features, the trap should be baited with food attractive to the rat. This ensures maximum results.

The chief *chemical control* is by poisoning, either through a single dose (acute poison) or a multiple dose (chronic poison).

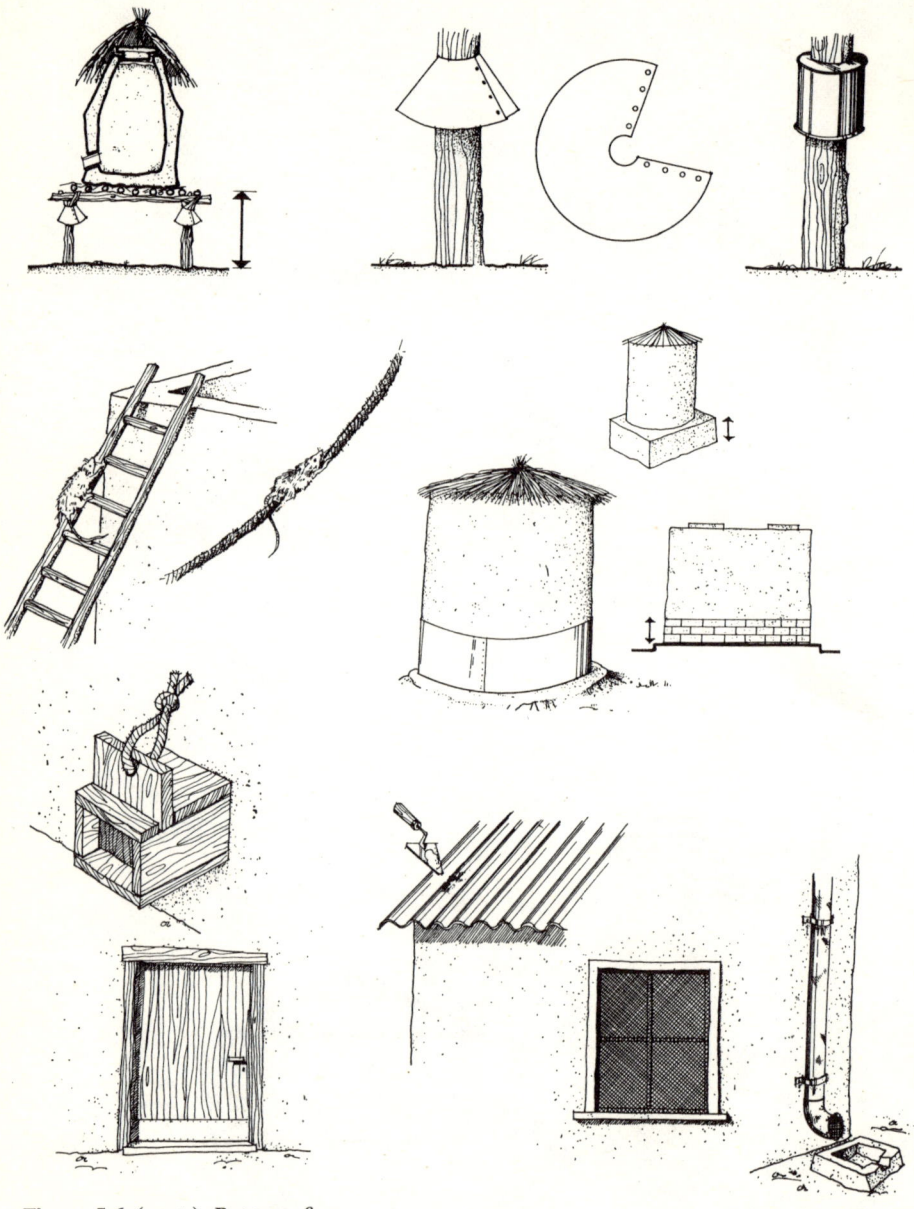

Figure 5.6 (cont.) *Rat proofing*

Single dose. Zinc phosphide is the most widely used. Two stages are essential for effective control.
(a) Pre-baiting. The sites, baits and containers should be the same as those to be used for the poison at the next stage. The more attractive the bait, the

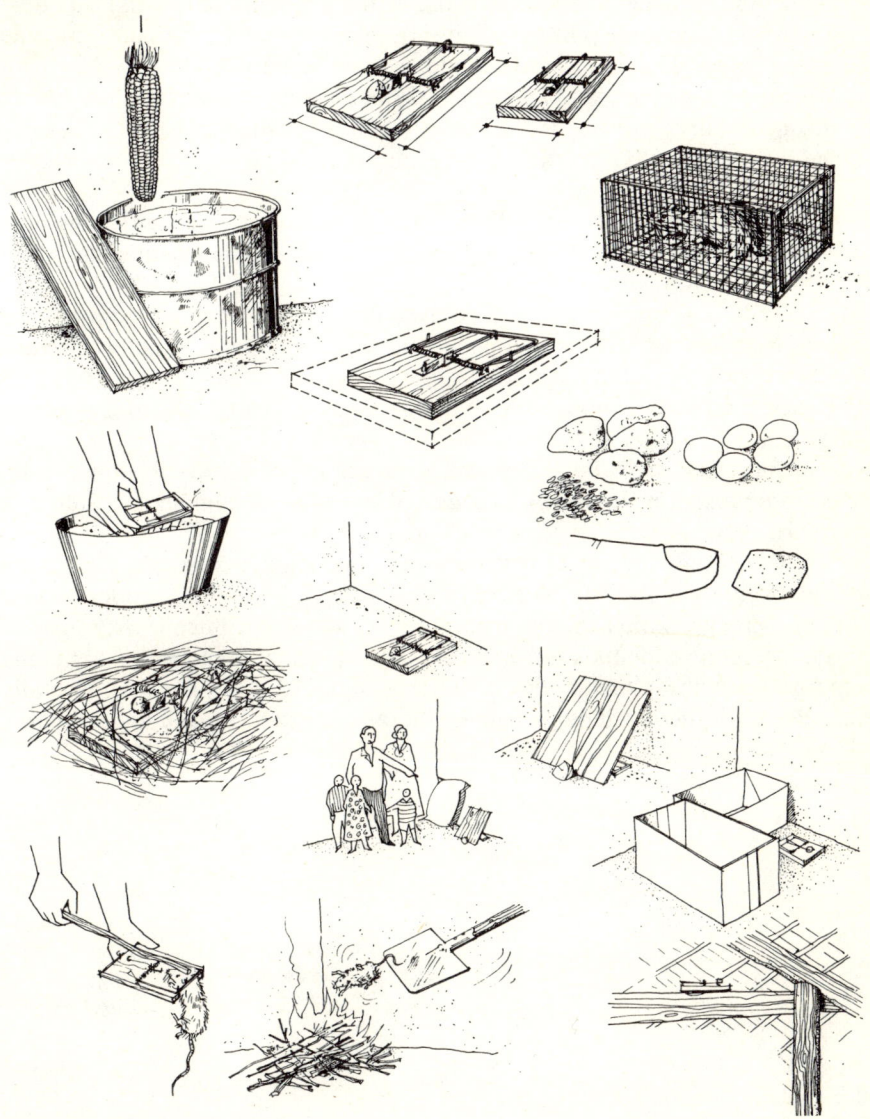

Figure 5.7 *Mechanical control of rats*

more successful the control. Cooked rice, soaked wheat or maize, and flour mixed with syrup are attractive baits. Pre-baiting should continue for three to four days; freshly prepared bait should be provided each day.

(b) Baiting with poison. One part of zinc phosphide is mixed evenly with 20 to 40 parts of a bait similar to that used for pre-baiting. The special containers used for pre-baiting should then be furnished with the poisoned bait and left at sunset in the same positions as those of the pre-bait containers. The next morning the remaining poisoned bait should be removed from the containers and destroyed. The containers should then be replaced, after loading with pre-baiting (non-poisonous) material. If this is eaten it indicates that further control measures are needed and the whole operation must be repeated. Dead rodents should be removed every day.

Multi-dose chronic poisons. These are generally blood anticoagulants causing death by internal bleeding. Their main advantages over single-dose poisons are as follows:

(a) rat colonies are not alarmed because deaths appear to be from natural causes, and they will continue to ingest the poisoned bait, ultimately giving a better overall control than single-dose poisons;

(b) such poisons do not give rise to bait-shyness, and no pre-baiting is necessary; and

(c) such poisons are used in very small quantities, and they are slow-acting, therefore presenting less risk of accidental ingestion by humans and domestic animals.

The manufacturers' instructions for anticoagulants should always be closely followed, and the bait containers located where only rodents have access. Rats are killed in about ten days, although for mice it may take 20 days. Affected rodents seek fresh air and water and therefore generally emerge from the store to die. Carcasses should be disposed of carefully because any anticoagulants remaining therein will affect scavenging animals.

6. Drying

6.1 Introduction

Cereals are annual crops grown for their edible starchy seeds. They are humans' main source of carbohydrate. The natural sequence of events is that the grains mature, ripen, become dormant and are then scattered on the ground where they eventually germinate and produce a new plant.

This sequence of events is successfully interrupted when the seeds are harvested, dried and stored. Their metabolic activity is thus reduced to such a low level that they do not deteriorate significantly; but the grains have to be protected against losses which may arise from other causes.

Temperature and *moisture content* are the two factors we have learned to use to control deterioration during storage. Combinations of these factors which have proved to be effective for safe storage against specified agents are shown in Figure 6.1. The total combined effect of these factors is shown in Figure 6.2.

In the tropics, temperature control is too costly, so that drying remains the most cost-effective method. Even at low moisture levels, however, stored grain is not safe from deterioration caused by insects when the temperature is above 15°C.

Many different methods and systems are used for drying and storing grain. To evaluate them for the farmer, village or cooperative, the basic principles involved in the exchange of moisture between the air and grains must be understood.

6.2 Air and water vapour: psychrometry

Atmospheric air consists of *dry air* and *water vapour*. For example, in Ibadan, Nigeria, at noon on a day in mid-October, 1 m^3 of atmosphere contained 1 131.2 g dry air and 20.36 g water vapour. The dry air and the water vapour were evenly distributed throughout the 1-m^3 space. The temperature was 30°C and the relative humidity was 66 percent.

The quantity of water vapour present is usually expressed as a proportion of the dry air present in the same volume as the water vapour, and is known as the specific humidity (s). In this case, the specific humidity was as follows:

$$\text{Specific humidity (s)} = \frac{20.36 \text{ g water}}{1\ 131.20 \text{ g dry air}} = 0.018 \text{ kg water/kg air}$$

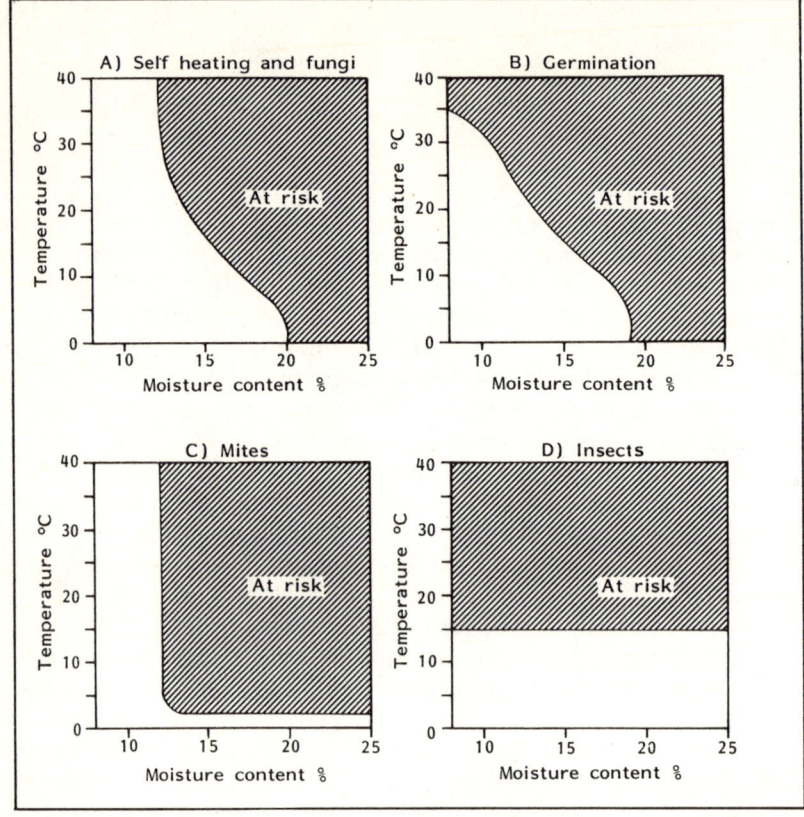

Figure 6.1 *Combinations of moisture content and temperature providing safe storage conditions for bulk-stored grains.* Source: *K.A. McClean.* Drying and storing combinable crops. *Farming Press Ltd, 1980.*

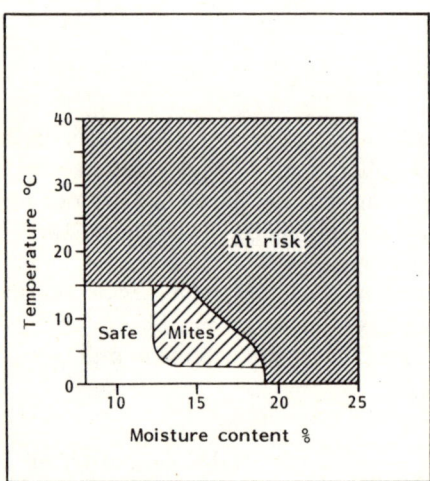

Figure 6.2 *Combinations of moisture content and temperature providing safe storage of grain against the four main agencies.* Note: *Area at risk from mites is shown separately because mites are not always a problem*

TABLE 2. **Air and water vapour in 1 m^3 of space**

Measurement	Condition[a]		
	1	2	3
Water vapour wt (g)	20.36	20.36	20.36
Dry air wt (g)	1 130	1 130	1 130
Dry bulb temp. (°C)	30.0	35.0	24.0
Wet bulb temp. (°C)	25.1	26.2	23.6
Specific humidity	0.018	0.018	0.018
Relative humidity (%)	66	49	95
Dew point (°C)	23.3	23.3	23

[a] See Figure 6.4.

No rain fell on that day, and by 15.00 h the air temperature had risen to 35°C. The specific humidity was still 0.018 because there had been no change in the amount of water vapour, but the relative humidity had fallen to 49 percent. (The small change in density of the air/water vapour mixture caused by these changes is ignored.) Was it possible that the air was now drier?

In effect, the *relative humidity* (rh) measures the percentage saturation of the space by water vapour; and because warm air can coexist in the same space with far more water vapour than cold air (see Table 3), the rh (or percentage saturation) has fallen, although no water vapour has been removed from, or added to, the space. The temperature only had risen. The space is capable of holding far more water vapour at 35°C (see Table 3).

By 22.00 h the air temperature had fallen to 24°C and the *relative* humidity had risen to 95 percent, although the specific humidity remained at 0.018 kg water/kg dry air.

The relationships between temperature and humidity are complex but may be represented by mathematical equations, which allow the effects of change in any of the factors to be calculated, or by a *psychrometric chart*. This chart is in the form of a graph, the axes of which are temperature (horizontal) and specific humidity (vertical). Values of relative humidity are represented by a series of curved lines. A simplified form of the chart is shown in Figure 6.4 where the three points representing the situations at noon, 15.00 and 22.00 h are plotted, as already discussed. Using the chart, refer back to Section 6.2 and check the values of relative humidity, specific humidity and temperature at each point.

The fourth scale of figures on the chart represents the wet-bulb temperature and is used to determine the working point on the chart, because it is easier to measure temperature than the other factors involved.

TABLE 3. **Water vapour carrying capacity of atmospheric space at various temperatures and normal pressure**

Temperature (°C)	Specific humidity (kg/kg)	Percentage of value (at 20° C)
0	0.0038	26
10	0.0076	51
15	0.0107	72
20	0.0148	100
25	0.0202	136
30	0.0274	185
35	0.038	257
40	0.050	338
50	0.083	561
60	0.150	1 014
70	0.330	2 230

If the space is not saturated then it can, of course, accept more water vapour. If a piece of clean cotton cloth is made wet with distilled water and air is forced to move past it, some water will evaporate from the cloth. This evaporation requires energy which is absorbed from the wet cloth, resulting in a fall in its temperature. Equilibrium is soon reached when the temperature of the cloth remains steady. This is the *wet-bulb temperature*, a term derived

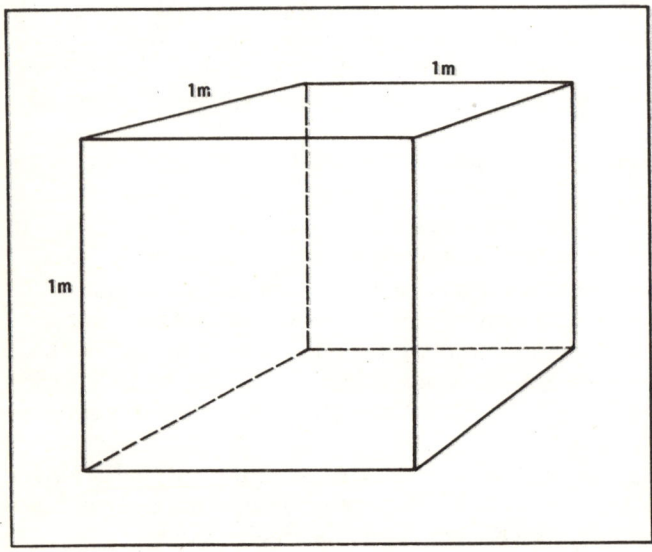

Figure 6.3
Air and water vapour in 1 m^3

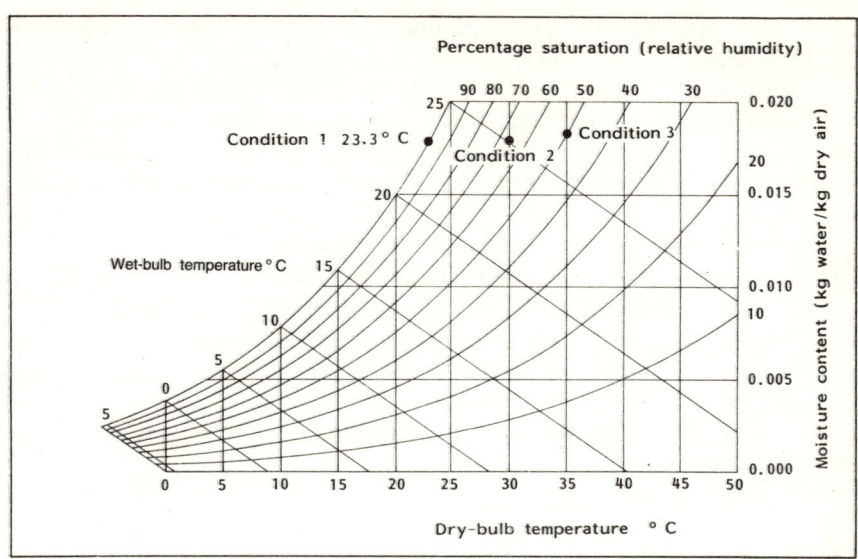

Figure 6.4 *Psychrometric chart.* Source: *Chartered Institute of Building Services (London, UK)*

TABLE 4. **Weight of water lost from wet grain when dried (g/kg)**

Initial moisture content (%)	Final moisture content (%)									
	19	18	17	16	15	14	13	12	11	10
30	136	146	157	167	176	186	195	205	213	222
20	125	134	145	155	165	174	184	193	202	211
28	111	122	133	143	153	163	172	182	191	200
27	99	110	120	131	141	151	161	170	180	189
26	86	98	108	119	129	140	149	159	169	178
25	74	85	96	107	118	128	138	148	157	167
24	62	73	84	95	106	116	126	136	146	156
23	49	61	72	83	94	105	115	125	135	145
22	37	49	60	71	82	93	103	114	124	133
21	25	37	48	60	71	81	92	102	112	122
20	12	24	36	48	59	70	80	91	101	111
19		12	24	36	47	58	69	80	90	100
18			12	24	35	47	57	68	79	89
17				12	24	35	46	57	67	78
16					12	23	35	45	56	67
15						12	23	34	45	56

from the practice of enclosing a thermometer bulb with the piece of wet cloth. Dry-bulb temperatures and wet-bulb temperatures are read from thermometers mounted side by side in a whirling hygrometer. From these two readings, all other characteristics of the air/water relationship may be established from the chart. The wet-bulb depression is the difference between wet-bulb and dry-bulb temperatures and may be used to find relative humidity from tables.

The important points are summarized as follows.
(a) Ambient air consists of dry air and water vapour.
(b) The ratio of water vapour to dry air is very small.
(c) If the *actual* water content of the space occupied by ambient air does not change, then relative humidity is reduced as temperature is raised; and is increased as temperature is lowered.
(d) These relationships can be represented by a graph known as the psychrometric chart.

6.3 Moisture content and relative humidity

The moisture content of cereal grains exposed to ambient air changes continuously in response to the relative humidity (rh) of the air. The greater the rh, the greater the moisture content of the grain. The changes take place relatively slowly, but given sufficient time a new value of relative humidity is maintained and a *near-equilibrium value of moisture content* is reached.

This characteristic of an *equilibrium* (equilibrium moisture content referring to grain and equilibrium relative humidity referring to the air surrounding the grain) is extremely useful because it can be applied in practice to adjust the moisture content of grain during drying and storage.

The equilibrium moisture contents for a wide range of grains have been measured, some of which are given in Table 5. These have been determined by exposing grains to atmospheres of different relative humidities and

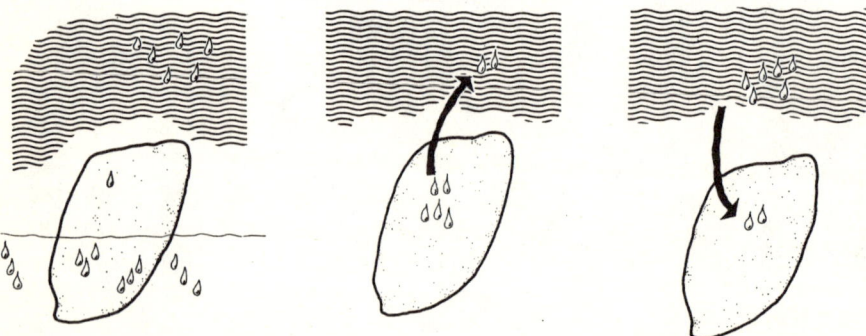

Figure 6.5 *Water movement*

TABLE 5. **Equilibrium moisture contents of various seeds**

Seed	Air relative humidity (%)						
	40	50	60	70	75	80	90
Wheat	10.7	12.0	13.7	15.6	16.6	17.6	23.0
Maize	11.0	12.0	13.0	15.0	15.5	16.0	20.0
Rye	10.0	11.6	13.2	14.8	16.1	17.3	24.6
Peas	9.4	11.1	13.1	15.5	17.2	19.5	27.7
Beans	9.1	11.1	13.1	15.8	18.0	20.4	28.0
Grass	8.9	10.3	11.6	13.9	15.4	17.4	23.3
Onions	8.3	9.6	10.8	12.6	14.1	16.2	23.5

Source: K.A. McLean. *ASAE Year Book*.

measuring the moisture content of the grain after several weeks of exposure. Obviously, many other factors are involved in determining a grain's equilibrium value, but these tabulated values serve as a very useful guide.

6.4 Drying

Drying involves transferring water from inside the grain to the grain's surface, transforming it to water vapour and then dispersing the water vapour into the atmosphere.

In order to achieve drying, the following three resources must be available:
- a source of water in the grain
- a source of energy to convert the liquid water into water vapour
- a sink, or destination, to receive the water vapour, and remove it from the grain mass

6.4.1 **Source of water.** This is the excess water in the grain which, if allowed to remain, will lead to deterioration. The quantity of excess water to be removed in drying grain from one moisture content to another can be found in Table 4. This is not a food loss.

6.4.2 **Sources of energy.** Energy is used to evaporate the water, i.e. to convert liquid water to water vapour. Twice as much energy is needed to evaporate water from grain as that required to evaporate boiling water in a pan. Water will evaporate at any temperature below boiling point, and still require about the same amount of energy to convert into water vapour as when it is boiling. The main sources of energy are solar energy, used either directly as in sun-drying, or indirectly as in crib drying; and energy derived from burning material such as wood, coal, gas or oil. In all cases, except direct sun-drying, the energy is transferred to air (the temperature of which is thus increased), and thereafter to the grain where it evaporates the water and is itself cooled.

During the day ambient air increases in temperature as a direct result of the sun's heating effect. The air stores the sun's energy which can then be used to dry grain during the middle of the day. The effect of increasing air temperature on relative humidity has been shown in Section 6.2. Thus, during the middle hours of the day, the air temperature is raised but relative humidity is also reduced. This air therefore has an increased capacity for drying through its increased temperature, and therefore more energy available for evaporation. Because its relative humidity is reduced, it can also absorb more water vapour. The combined effect of increased temperature and lower humidity, and the interrelationship of the two, are shown in Figure 6.6.

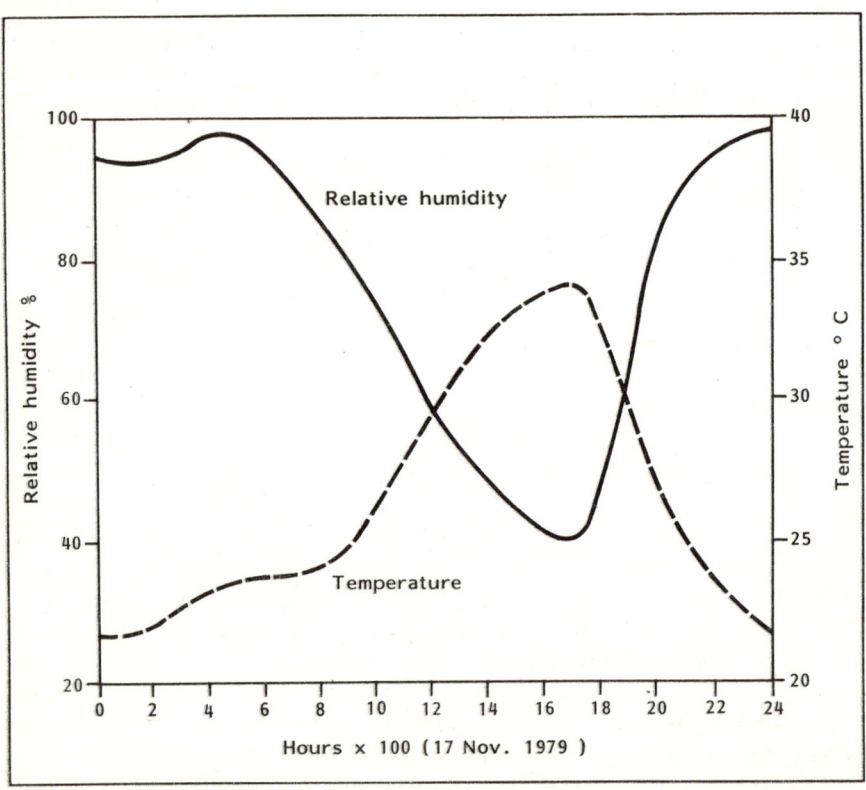

Figure 6.6 *Typical daily variation of air temperature and relative humidity (Ibadan, Nigeria)*

6.4.3 **Vapour sink.** The evaporated water must be removed from the vicinity of the grain. If it is allowed to remain the relative humidity, as already noted, is increased, so that evaporation ceases even if the grain is wet and energy is available. In practice, the air surrounding the grain is artificially replaced by new air through diffusion (as in cribs), or by forcing it away using a fan (as in a batch or a continuous flow drier) or relying on convection (as in the Brook drier).

Diffusion of air in still conditions is a very slow process. This is one reason for recommending a very narrow crib in difficult drying environments.

6.5 Types of driers

The designs of driers and their methods of construction offer a wide range of possible combinations to meet the requirements discussed in Section 6.4. A drier is useful only if it proves technically efficient at a low cost. At the farm and village level this means making the fullest use of locally available materials and expertise. A drier may be excellent from a technical viewpoint, but if it is too costly it will be of no more use than a low-cost drier that fails to conform to the fundamental principles of drying.

There are many types of drier and many models of each type. When choosing a drier, it is important to remember that it must:
(a) be acceptable to the intended user;
(b) conform to the fundamental principles of drying; and
(c) be low-cost compared to previous drying methods so that the user can quickly benefit from it.

Drying may be an operation separate from storage; or drying and storage can be combined.

6.5.1 **Drying as an operation separate from storage.** This type of drying operation is separate from any storage operations which may follow it. It is important because the grain is mixed after it has been dried. Kernels that are too dry will absorb moisture from kernels that are too wet. When they are mixed after drying, the desirable average final moisture content will be achieved. In *storage-drying* operations, no mixing occurs because the crop is already in store. No differences in moisture content should exist after drying; there must be no over-drying or under-drying of individual kernels.

In *sun-drying*, or exposing grain to the sun, the only requirements are a level surface and sufficient labour to spread the material, turn it and collect it in case of rain or when the grain is dry. The cheapest drying surface is tamped earth, but this has the disadvantage of contaminating the grain and subjecting it to ground moisture. For small quantities a tarpaulin or plastic sheet prevents these difficulties, and the edges can be turned in to cover the grain when rain threatens or at night. Black plastic sheeting with grain spread to 40 mm

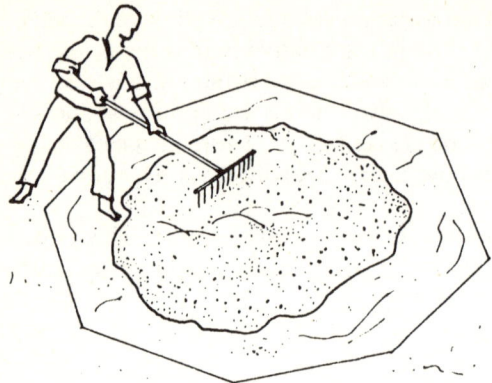

Figure 6.7 *Drying as a separate operation from storage*

and frequently turned gives the fastest drying. Labour requirements are, however, a major constraint.

Using a *convection drier with added heat* is another drying method. The Brook drier (see Fig. 6.9) is an example of a convection drier. The crop is spread in a thin layer on a perforated floor. The floor forms the top of a plenum chamber in which there is a source of heat. The perforated floor has

Figure 6.8 *Combined drying and storage*

sides which extend upward as high as possible, since the crop has to be loaded and unloaded over the sides unless they are removable (which adds to the cost). Air enters the plenum chamber through holes in its sides and end, is heated and rises through the perforated floor and the product to be dried by natural convection.

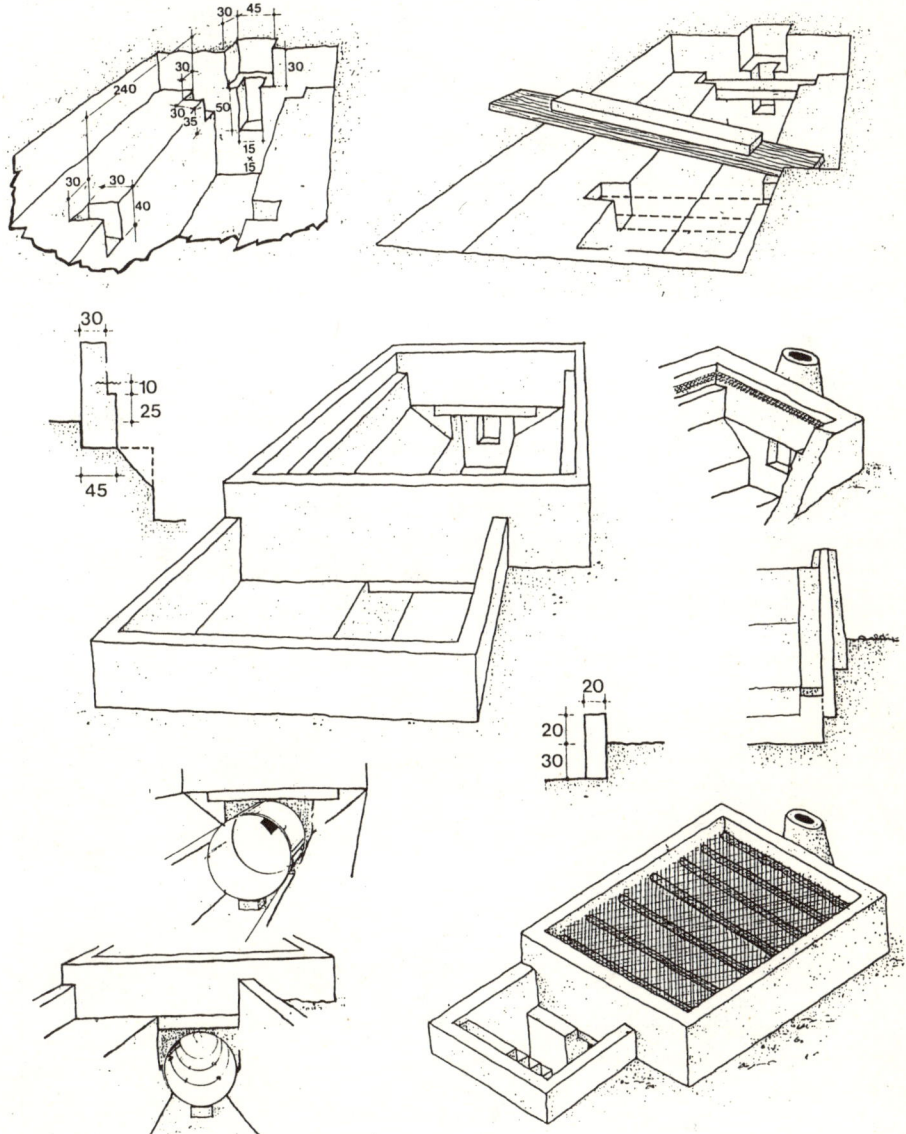

Figure 6.9 *Natural convection drier, general view*

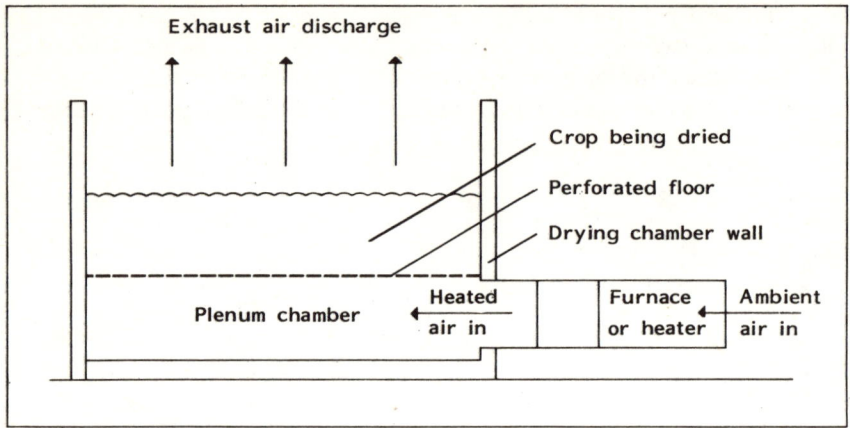

Figure 6.10 *Force-ventilated batch drier*

Another form of drier is the *force-ventilated batch drier*. In its simplest form this consists of a perforated floor overlying a plenum chamber. A fan passes air through the crop spread on the perforated floor. Normally the air is heated before being passed through the crop. After drying, the crop is removed, cooled and stored. This system dries crops quickly but is expensive to construct. It also requires components which are not readily available to the farmer. Moreover, it involves expenditures on fuel to drive the fan and maintenance of the equipment. There are many examples of this type of drier; a typical drier is illustrated in Figure 6.10.

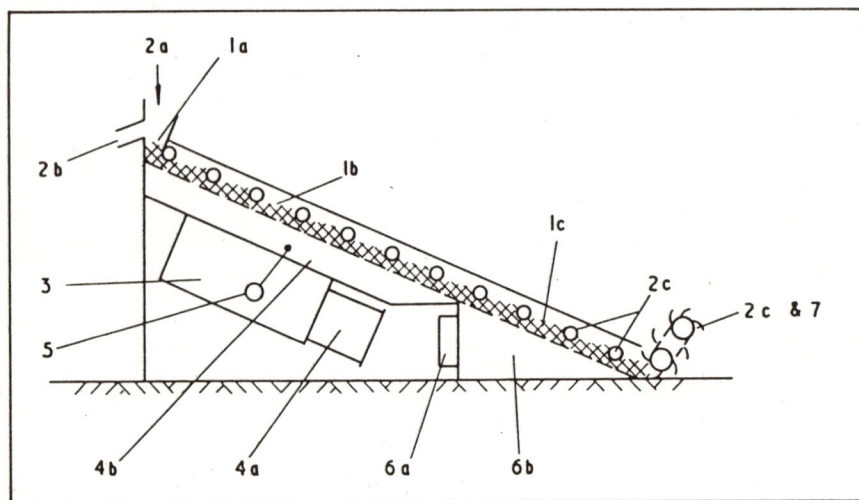

Figure 6.11 *Cascade drier*

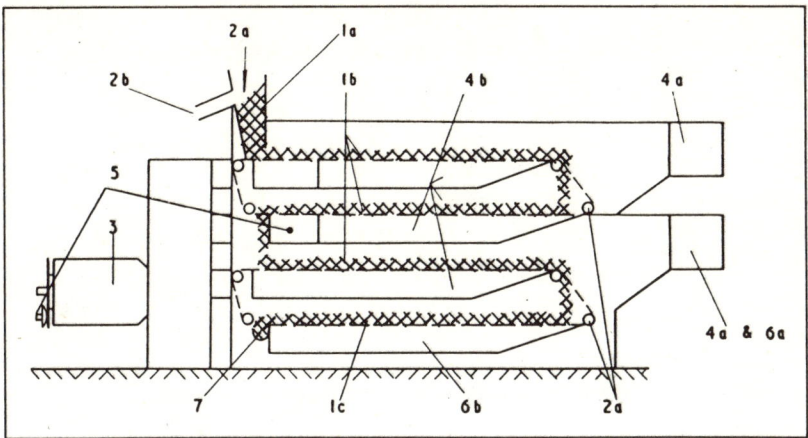

Figure 6.12 *Multiple-tier conveyor drier*

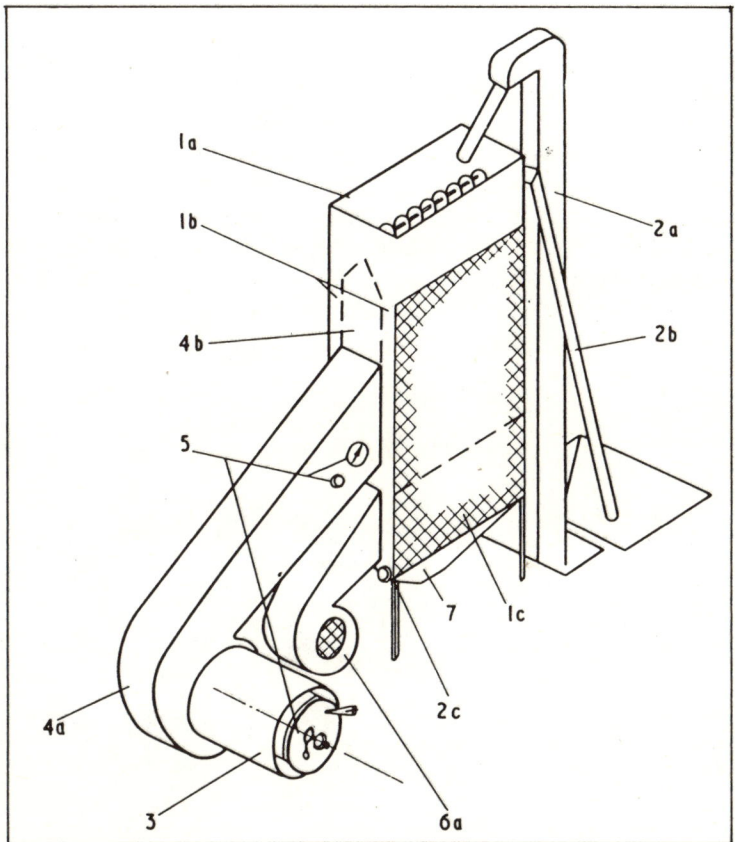

Figure 6.13 *Double-flow continuous drier*

Legend for Figs 6.11, 6.12 and 6.13

1. Grain compartments
    (a) reservoir
    (b) drying compartment(s)
    (c) cooling compartment(s)
2. Means of causing and regulating grain flow
    (a) feed
    (b) overflow
    (c) regulating mechanism
3. Air heater
4a. Hot-air fan(s)
 b. Hot-air chambers
5. Devices for controlling (and indicating) hot-air temperatures
6a. Cooling air fan
 b. Cooling air chambers
7. Grain discharge

In the *force-ventilated continuous flow drier,* the grain is continually moving through the drier as heated air is forced through it. It may move by gravity between two perforated walls or down a sloping perforated floor. In some driers the grain is moved over a horizontal perforated floor by slow-moving scrapers. The temperature of the air must be very accurately controlled to avoid damaging the grain. Specially designed furnaces and expensive fans are needed for this type of drier, which are suitable only for large-scale operations with a high throughput of grain. Examples of continuous flow driers are given in Figures 6.11, 6.12 and 6.13.

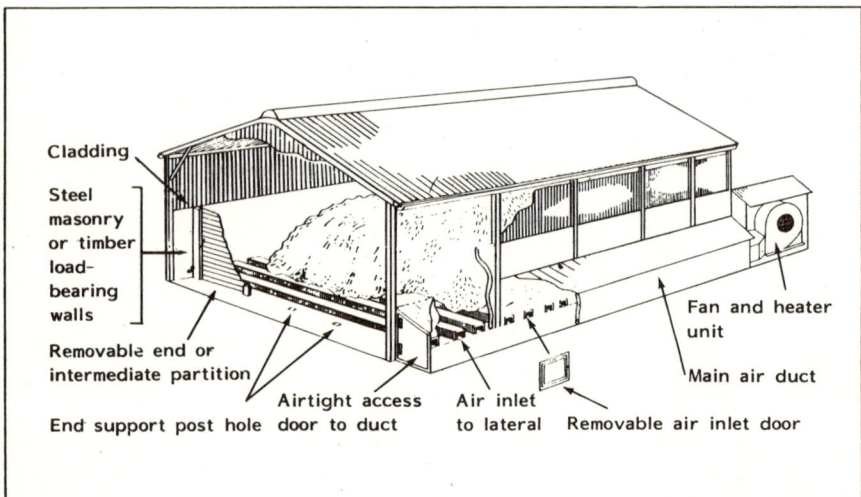

Figure 6.14 *On-floor storage drier*

6.5.2 **Drying combined with storage.** In this group, the same structure is used for storage and for drying. The storage structure is designed to allow the crop to dry during the early part of the storage period. In most storage driers the crop is not removed from the structure until it is needed. In some, however, the structure is used for batch drying.

One such drier is the *natural ventilation storage drier*. The most common example of this group is the crib, which is fully discussed in Section 10.

Forced ventilation storage driers can be ventilated bins and on-floor storage-drying systems. Both are used for bulk grain in large quantities. Considerable skill and experience are required to operate these systems successfully. Examples of this type of drier are shown in Figures 6.14 and 6.15.

Figure 6.11 shows the *continuous flow louvred-bed grain drier* that runs on the cascade principle. This is a gravity-flow drier assisted by the downdraught from the louvred bed and with depth controlled by a series of roller dams. The moisture extraction rate is controlled by a variable speed drive to the output elevator, lifting grain evenly from the full width of the bed at the base of the cooling section. The machine is suitable for all free-flowing grain and most granular material. Being almost entirely self-cleaning, it is a popular choice of seedsmen. The cascade drier is particularly suitable for drying peas, beans, coffee and rice.

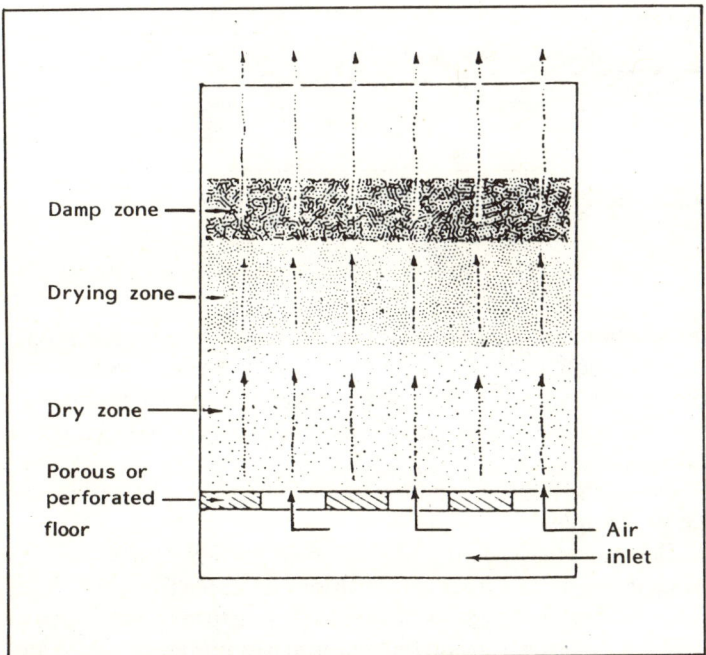

Figure 6.15 *Drying zones in bottom-ventilated storage bins*

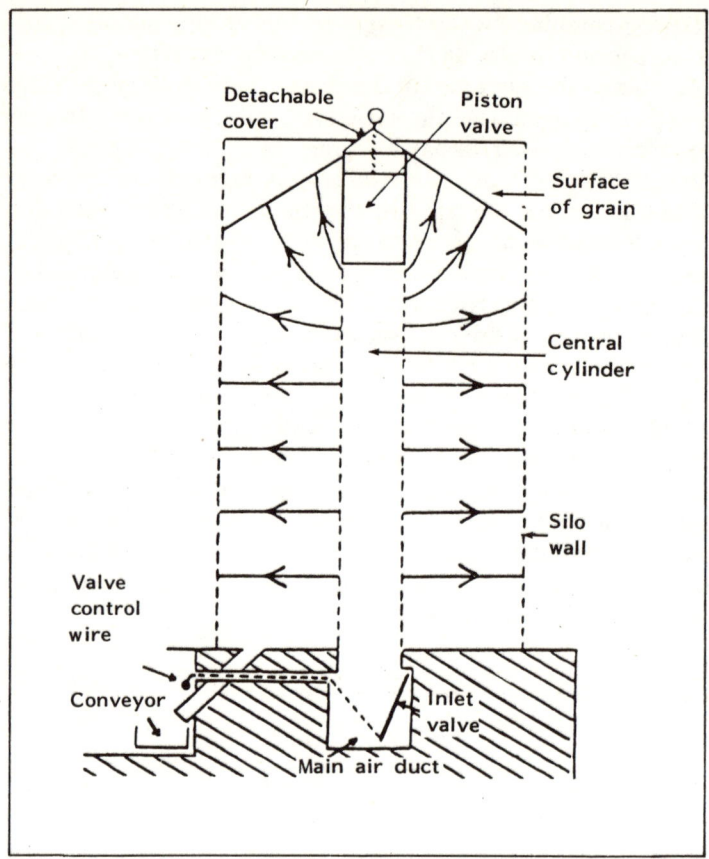

Figure 6.16 *Radially ventilated storage bin with central cylinder and perforated walls*

Most models are delivered in one piece, which makes installation simple and rapid. Standard models are available to give outputs from 2.5 to 12.5 tonnes per hour. The smaller cascade driers can be supplied as mobile units. Multiple-tier cascade driers (see Fig. 6.12) are manufactured to special order when site conditions dictate.

Figure 6.13 shows the continuous flow louvred-bed grain drier running on the double flow principle.

The flow of this machine is gently assisted by a variable speed, heavy-duty, roller-chain conveyor. The double flow drier is suitable for most grains and granular materials, including grains of very high moisture content such as rice and maize. Half-way along its route the material is completely turned and mixed as it falls from the uppermost to the lower bed. Advantage may be taken of the strong cleaning effect produced by the current of air passing through the curtain of grain at this point.

The feed and discharge points of this model are found at one end, which simplifies installation. The remainder of the drier is usually positioned outside the building to economize on space, and to ensure that moist air and dust are exhausted into the atmosphere. Roof supports are supplied to take six big asbestos sheets. Standard models are available to give outputs from 4 to 85 tonnes per hour. Mobile double flow driers can be supplied to give outputs up to 21 tonnes per hour. Multi-flow driers may be supplied for special applications.

7. Warehouses

Warehouses are durable, general-purpose storage structures, providing secure protection from rain, sun and wind.

Grain storage warehouses are specialized structures designed specifically for storing cereal grains and legumes. Grains may be stored in bulk or in bags. Most grain storage warehouses are designed for bag storage. Capacities range from 50 to 5 000 tonnes of bagged grain per warehouse, requiring from 50- to 2 000-m^2 floor area. Smaller sizes for storing between 5 and 10 tonnes of grain in bags are common, but storage costs per tonne then tend to be high, and alternative methods such as bin storage should be investigated.

7.1 Warehouse construction

A general warehouse design consists of a concrete frame with concrete block walls; a metal roof truss; a galvanized, corrugated iron roof covering; and a floor area of 600 m^2, with a 15-m width and 40-m length. This type of structure is standard for the commercial building contractor; alternative designs are usually quoted at a higher unit price.

For warehouses smaller than 200 m^2 locally available material, such as mud block and timber, may be used. While not ideal for long-term storage, they provide adequate short-term storage at, for example, primary procurement centres.

7.2 Cost of construction

There is a marked economy of scale in warehouse construction, ranging from US$ 200 per m^2 for a 500-m^2 floor area, to US$ 180 per m^2 for a 2 000-m^2 floor area in the major cities. In remote areas basic costs are as much as 50 percent higher and there may be up to 30 percent additional costs to provide road access and services.

Thus, in remote areas, the price per square metre compared with that in areas close to cities may often be 30 percent greater for small stores and as much as 40 percent greater for floor areas up to 1 000 m^2.

Figure 7.1 *Warehouses*

7.3 Usable volume

Warehouses cannot be completely filled with bagged grain. Access ways are required, and the apparent wasted space taken up by smaller gangways, headroom and areas around stacks is essential for ventilation, access, hygiene and fumigation. For a 500-tonne store the usable volume can be less than 50 percent of the gross internal volume below the level of the eaves. As store size increases the usable volume increases to a maximum of about 80 percent for 10 000-tonne stores. The usable volume is also decreased, for a given size of warehouse, by (a) an increased range of products stored; (b) short-term storage where stacks are continually being broken and rebuilt; (c) pest infestation and/or (d) poor management.

7.4 Care of produce in warehouses

(a) *Prevent damp from the floor reaching the produce*

During construction a water barrier or membrane can be included in the concrete floor of the warehouse. Pallets and dunnage are used to form a barrier against damp.

(b) *Prevent damp from walls reaching the produce*

Figure 7.2 *Dunnage*

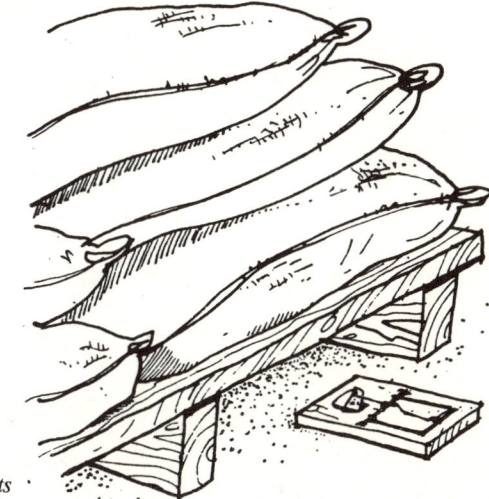

Figure 7.3 *Pallets*

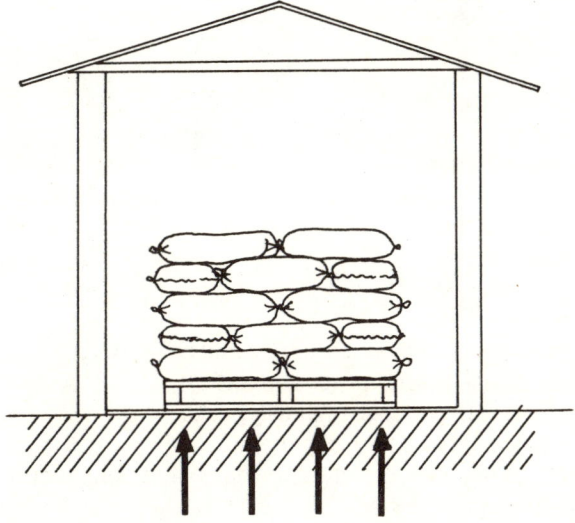

Figure 7.4 *Space between produce and walls*

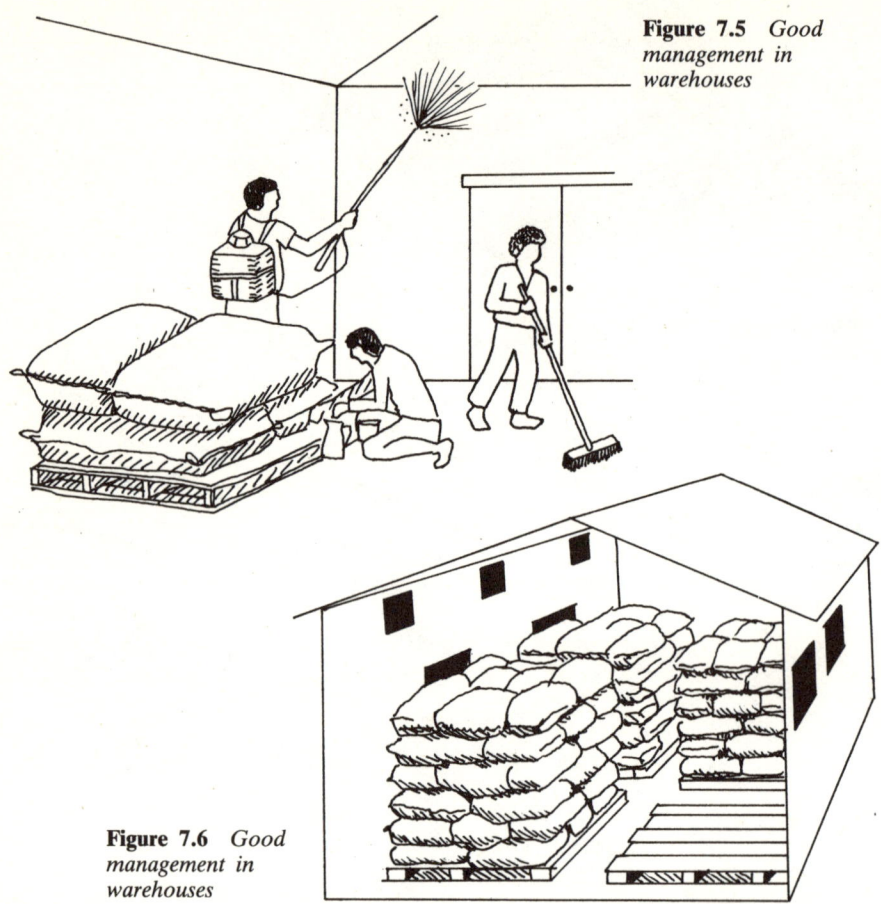

Figure 7.5 *Good management in warehouses*

Figure 7.6 *Good management in warehouses*

(c) *Stack the sacks properly to allow:*
    (i) optimal use of space;
    (ii) ease of sweeping the floors;
    (iii) ease of inspection of produce for the presence of rodents and insects;
    (iv) ease of counting sacks; and
    (v) ventilation of sacks.

(d) *Control insects and rodents by:*
    (i) closing all holes in doors, roofs, etc. where pests can enter;
    (ii) repairing cracks in walls where pests can hide;
    (iii) treating the building and produce against pests;
    (iv) keeping the warehouse absolutely clean; and
    (v) removing and destroying any infected residues that might contaminate newly introduced pests.

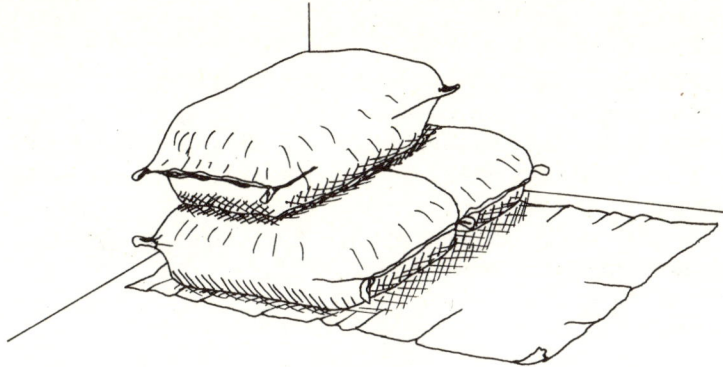

Figure 7.7 *Dunnage: waterproof sheet*

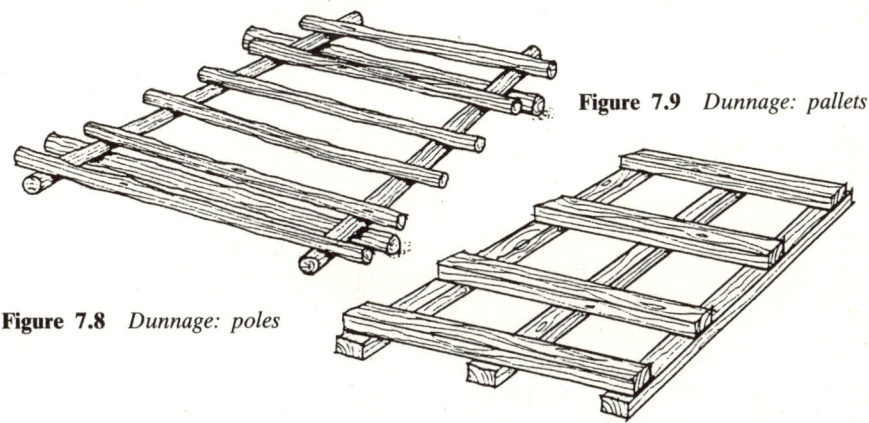

Figure 7.9 *Dunnage: pallets*

Figure 7.8 *Dunnage: poles*

7.5 Dunnage

Dunnage is material that can be placed between the floor of a warehouse and the sacks of produce to prevent moisture moving from the floor into the produce, and thereby causing moulding and rotting.

The simplest form of dunnage is simply a thick waterproof mat or unpunctured plastic sheet, on which the sacks are placed.

Alternatively, straight poles are laid on the floor and the sacks are placed on top.

The more expensive type of dunnage consists of two layers of planks, securely fastened to cross members to keep them separated. If made from sawn timber they are known as pallets and are suitable for handling by forklift trucks. Pallets should be inspected and sprayed before use to avoid cross-infestation and damage to sacks by protruding nails and split wood.

7.6 Stacking of sacks

If sacks are laid on top of one another, with no overlap in successive layers, the stack will be very unstable. The recommended alternative is to "tie" successive layers by arranging the sacks differently in each layer. This not only produces a safer stack, it also makes stock-taking easier as the bags can quickly be counted.

Sacks are generally stacked in layers of three, five or eight per layer. Odd and even layers are arranged as shown in Figure 7.10.

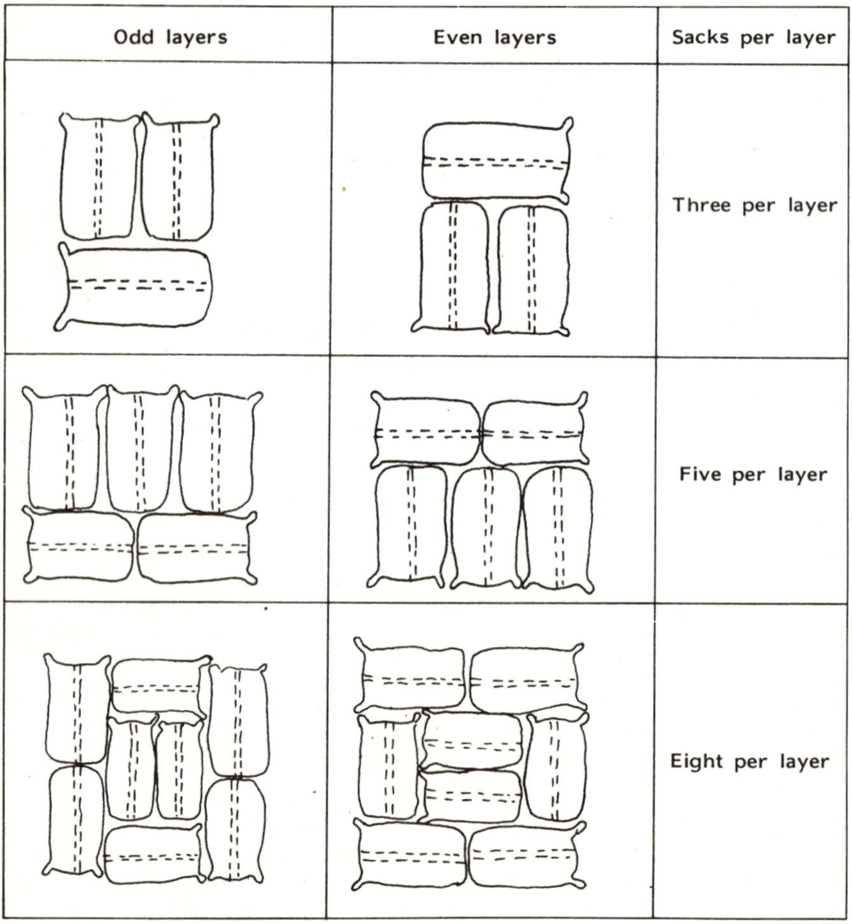

Odd layers	Even layers	Sacks per layer
		Three per layer
		Five per layer
		Eight per layer

Figure 7.10 *Stacking sacks*

7.7 Insect control in sacks stored in warehouses

There are three common chemical methods for controlling insects in sacks stored in a warehouse:
(a) admixture of insecticidal dusts with the produce before loading it into the sack;
(b) the spraying of successive layers of sacks with liquid insecticides or dusts as the stack is built; and
(c) enclosing a fumigant with the sacks under a gasproof sheet.

The admixture of insecticidal dusts can be very effective if a suitable insecticide is used. Recently, some synthetic pyrethroids and pirimiphos-methyl dust, applied at rates of between 2.5 ppm and 15 ppm active ingredient (depending on the insecticide), have been found to completely eliminate insects in stored bags for at least eight months.

Mixing of the dusts with the grain can be done in various ways, such as shovel mixing on a tarpaulin, or, for large-scale operations, a drum with an eccentric axle is used.

The admixture of dusts with stored grain gives rise to a potential health hazard and is not to be recommended, unless a very safe insecticide is used and the grain is only to be consumed after a prolonged period in storage.

Spraying or dusting successive layers of sacks with insecticides (as shown) is less hazardous to humans, but is not always effective. Recently, however, pirimiphos-methyl (as emulsifiable concentrate, Actellic 50 ec) was applied undiluted (50 ec) at the rate of two to three strokes per bag with a simple domestic applicator; and it largely eliminated weevils from heavily

Figure 7.11 *Drum mixer*

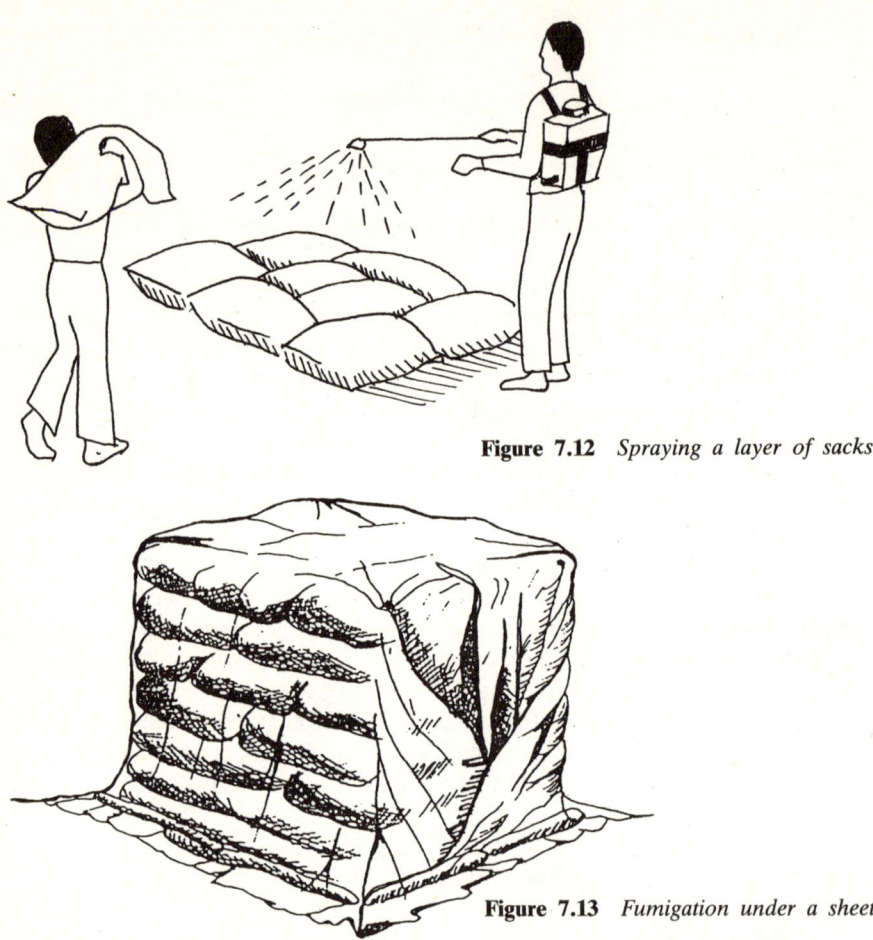

Figure 7.12 *Spraying a layer of sacks*

Figure 7.13 *Fumigation under a sheet*

infested sacks of maize and kept the population at a very low level even after eight months. But there is always a danger in applying undiluted insecticides.

Ultimately, the most satisfactory method of insect elimination and control in bagged grain is by fumigation. A gas is released among bags covered by a gasproof sheet, which is held down by "sand snakes" or a heavy chain wrapped in hessian (see Fig. 7.13). The sheeted stack is left for at least three days.

For relatively small-scale storage (100-300 tonnes) the most convenient fumigant to use is aluminium phosphide, which releases phosphine gas when it absorbs moisture. One tablet of the fumigant for every two bags is recommended, provided the stack is of a size that will be hermetically enclosed within two hours. Phosphine and other fumigant gases may also be used effectively for larger quantities of grain.

8. Central storage

In central storage and grain-handling facilities, there are the conflicting objectives of providing for large throughputs and for long-term reserve storage.

Port installations involved in either importing or exporting grain require mechanical handling facilities that can only be justified by a high degree of utilization. The benefits of being able to handle large volumes of bulk grain quickly lie in the prompt turnaround of ships, railway freight cars and road transport, thereby avoiding costly delays.

Such facilities are not required, however, for long-term storage where grain, in exteme cases, may remain for several years. It is always advisable that emergency grain reserves of this kind be cycled into the normal distribution system on the basis of a first-in first-out (FIFO) queueing system.

Central storage facilities are usually provided with drying facilities to treat incoming grain that is above the regulation 13-percent moisture content wet basis for longer-term storage.

9. Pest control in stored produce

9.1 Introduction

Various techniques are used to control insect pests in stored produce — from sunning and smoking on the traditional farm to irradiation in large-scale bulk handling. This section of the manual is concerned only with proved techniques suitable for use in small- and medium-scale storage under tropical conditions.

Special recommendations are difficult to make. A technique must be tested for a particular situation, since it may become inappropriate as a result of changes in:
(a) economics (the value of the product, relative to the cost of materials and labour);
(b) pest problems (such as occurrence and resistance); and
(c) techniques within the farming system or through the availability of new products.

Therefore, it is important to consider both:
(a) economics; and
(b) technical specifications for effectiveness against target pests and hazards to farmer and consumer.

Will the improvement resulting from the use of a control technique pay for the cost of carrying it out? This question can only be answered satisfactorily by field trials supported by an effective loss assessment.

9.2 Pest control techniques

9.2.1 **Sanitation.** It is crucially important to reduce the initial pest population and prevent development of any insect pests in the crop products. Before bringing a new crop into store, the following steps are necessary.
(a) *Remove infested material.* Do not mix new grain with old; old material that must be kept should be thoroughly fumigated.
(b) *Clean the storage structure:*
    (i) brush away all traces of spilled grain, dust, etc.;
    (ii) remove dust from handling equipment and machinery;
    (iii) disinfect sacks and baskets by sunning or chemical treatment; and
    (iv) be aware of the following specifications:

Figure 9.1 *Cleaning storage structure*

Figure 9.2 *Disinfecting sacks*

— large structures usually require chemical treatment; and
— small rural structures can be cleaned by using smoke, and making use of the sun and rain — after some time insects will usually leave a clean empty crib or rumbu.

Take control measures early to prevent infestation of crops maturing in the field.

9.2.2 **Natural resistance.** Crop varieties differ in their susceptibility to storage pests.

Traditional varieties are usually more resistant to storage pests than new varieties. If new varieties are introduced, measures must be taken to improve storage techniques and pest control.

Some new varieties of maize and cowpea are now being selected for improved storage resistance and they are now becoming commercially available.

There are some useful general resistant characteristics in the following crop varieties:
(a) maize: good husk cover can reduce field infestation, and storing in the husk reduces the rate of pest increase;
(b) sorghum: varieties where the glumes cover the grain tend to be more resistant before threshing;
(c) rice: paddy rice is considerably more resistant to pests than milled rice;
(d) cowpeas: intact dry pods provide some protection against bruchids; if fumigation or airtight storage is impracticable, cowpeas are better stored unthreshed;
(e) cereals: grain hardness affects resistance.

9.2.3 **Hermetic or sealed storage.** In airtight conditions, reduced oxygen and increased carbon dioxide will eventually arrest insect and mould development.

Grain for human consumption or seed must be dry; in damp grain bacterial and enzymatic action will continue, causing tainting and loss of viability.

Bagged material must be protected; if the seal is broken (by insects, rodents or careless handling) the grain is unprotected and unventilated, and losses may be severe.

A method that has been found satisfactory in North Nigeria (a dry area) is that of storing threshed cowpeas in sealed plastic bags with cotton liners; the cotton prevents emerging insects from perforating the plastic bag.

9.2.4 **Chemical control: overview.** Methods of using insecticides on stored products are:
- dusting
- spraying
- fumigation

Insecticides usually show some degree of toxicity to humans, domestic animals, poultry, etc. and must be used with caution. Be sure to:
(a) read the manufacturers' instructions;
(b) choose a chemical with low toxicity to mammals and birds;
(c) stay within the recommended dosage; and
(d) protect workers with careful instruction, constant supervision and provision of protective clothing.

Insecticides are usually specific, and do not kill all insects and mites; choose a chemical approved for use in stores and/or on stored products that has either a "broad spectrum" or specific toxicity to moths and beetles. Mites may require special treatment.

Insecticides tend to lose their persistence, or effectiveness, with:
- high humidity
- high temperatures
- sunlight
- time

Stored chemicals must be protected from these factors to ensure their continued effectiveness. In stored products, long-persistence insecticides give long protection against pests, but increase the risk to the consumer. Insecticides vary widely in their persistence. Choose one appropriate to the job: for example, persistent chemicals for the treatment of storage structures, non-persistent ones for space-spraying.

Insects can develop physiological and behavioural resistance to insecticides. Excessive or inappropriate use of chemicals will lead to the insects becoming resistant; therefore, use the right dose and use insecticides only when strictly necessary.

9.2.5 **Fumigation.** Chemicals used to attack insects through their respiratory system are known as fumigants.

Fumigants may be formulated as:
- liquids
  carbon bisulphide
  carbon tetrachloride
  ethylene dichloride
  ethylene dibromide
- gases
  hydrogen cyanide
  methyl bromide
  phosphine

The concentration of a fumigant is measured in mg/l of space occupied.

The CT product is the concentration of the fumigant multiplied by the time in hours that will give a 99-percent kill of the pest concerned. Table 6 gives details of some CT products for commonly used fumigants; it serves as a guide only to their effectiveness because other factors may be involved.

TABLE 6. **CT product for specified insect pests (in mg/l/hr)**

Chemical	Insect species			
	Rhizopertha	Sitophilus	Tribolium	Trogoderma
Carbon bisulphide	294	325	560	700
Carbon tetrachloride	—	4 500	2 000	—
Ethylene dichloride	636	1 200	365	2 080
Methyl bromide	0.60	1.0	0.50	331

Some characteristics of the more common fumigants follow.

Carbon bisulphide appears to be a reasonably good fumigant, but it is very inflammable; a spark caused by a nail hitting a stone will cause the fumigant to explode, and for this reason it is now rarely used.

Carbon tetrachloride seems to be a poor fumigant, but in fact it penetrates grain very well and is often mixed with other fumigants which do not penetrate in order to carry them through the produce. It is not inflammable, but over a period of time it may injure the livers of those who use it.

Ethylene dichloride is inflammable and does not penetrate well. Its use is not recommended.

Ethylene dibromide appears to be effective, but it is absorbed very rapidly; its penetration through the grain is therefore very poor.

Methyl bromide is an excellent fumigant, but it has no smell and is very poisonous. It can therefore **only** be used by **trained teams**. It is not generally available.

Phosphine is an excellent fumigant and fairly easy to use. It is used as a mixture of aluminium phosphide and ammonium carbamate. These are stable if kept in sealed containers, but when exposed to the air they take up water and release phosphine, ammonia and carbon dioxide. Phosphine normally contains impurities which make it spontaneously inflammable, but in the presence of ammonia and carbon dioxide it is safe. The chemicals are formulated so that there are about 30 minutes available to distribute the mixture before the gas is released. The gas has a strong and unpleasant smell and is therefore easy to detect. Phosphine is the only fumigant that will not interfere with germination if the grain is to be used for seed. The others may affect germination if exposure to the fumigant is excessive or repeated.

Fumigation in practice involves consideration of the scale of the operation, and safety.

In small-scale fumigation, grain and other products may be fumigated with carbon tetrachloride in a dustbin or 150-l drum. About 150 ml are poured over the surface of the grain, the lid is sealed with paper glued round the junction between lid and base, and the grain is left for 14 days. If it is to be used for seed it should be aired after fumigation; otherwise germination may be affected.

In larger-scale fumigation, phosphine fumigation may be carried out using formulations which consist of tablets, pellets or powder in envelopes. The makers issue instructions for the quantities to be used. Sacks of grain may be fumigated under plastic sheeting: the grain is stacked on a sheet of plastic, covered with another sheet, and after the fumigant is inserted the top and bottom sheets are rolled together. Sandbags are placed on the roll to give an airtight seal.

If it is not possible to put a covering sheet over the grain the building should be sealed thoroughly before fumigation. Alternatively, if feasible, a sheet should be put over the whole structure before fumigation.

Safety must be a primary consideration. All fumigants can kill people as well as insects, and some may cause serious disorders in humans who are exposed to low concentrations over a long time. Consequently, stocks of fumigant should not be kept in offices or stores where people are working.

Accidents can happen, so two people should always be present when fumigating. Fumigation should only be carried out by trained staff operating under proper supervision.

If phosphine is in use, one person should have a respirator with the correct canister. Phosphine fumigation is extremely easy to carry out, and increasing numbers are using it without a proper knowledge of the dangers. Some fumigants are supplied in containers that cannot be resealed, which can be dangerous if some of the material is left in an office or even a home. When the cover is lifted after a period of fumigation there will be a high concentration of the gas for a short time and this can be dangerous. Ideally, the material should only be handled by crop protection staff; but as these are so few and the results of fumigation can be so good, there is commercial pressure to make the products widely available. Since it seems inevitable that untrained people will use phosphine, the media (press, radio and television) should be prompted to give as much guidance as possible on its proper use.

9.2.6 **Biological pest control.** These methods have been effective in some situations. *Bacillus thuringiensis* has been used for controlling some species of insect pest in stored cereals. Cats provide effective control of small numbers of rodents in and around farmsteads, but they should not be used in warehouses.

9.3 Chemical control: specific methods

9.3.1 **Insecticidal dusts.** The method of dispersing dusts usually involves using an admixture of dilute dust at 2.5-15 ppm active ingredient, depending on the insecticide, at the time of loading/bagging.

Figure 9.3 *Mixing insecticidal powder with grain*

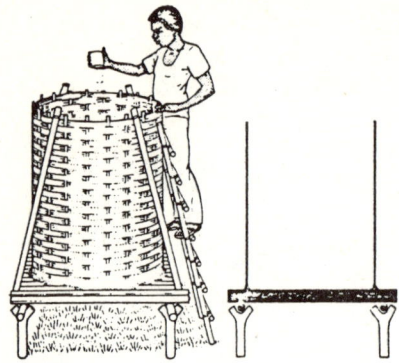
Sprinkle the inside walls and floor with a fine layer of insecticidal powder

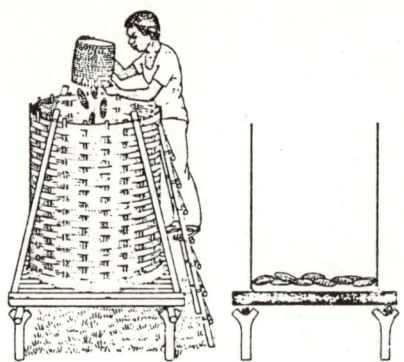
Completely cover the floor of the store with a layer of cobs

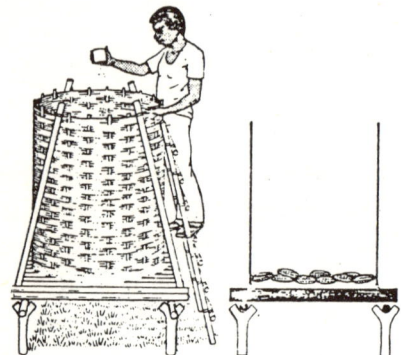
Sprinkle more insecticidal powder evenly over the layer of cobs

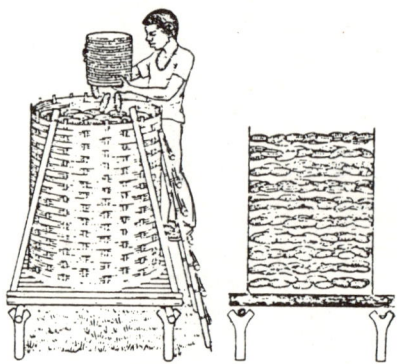
Put in more layers of cobs and sprinkle each layer with insecticidal powder

When you have put in all the cobs, cover the top layer with a thick coating of insecticidal powder

Figure 9.4 *Treating cob maize*

The appropriate quantity of dust is measured into a perforated tin or punctured plastic bag and sprinkled onto the produce layer by layer. For bulk grain the dust is mixed more effectively by shaking in a tin with the produce, shovelling on a groundsheet or mixing in a revolving drum. For larger-scale operations commercial devices are available.

There are several problems with this type of insecticide.

Applications of dusts include use in cribs and bulk stores but they are more effective in the latter. They are also only suitable for dry conditions.

(a) It is difficult to obtain accurate dosage and a thorough admixture.
(b) Application can only be carried out at loading, when it makes an extra demand on labour; so it is often not properly done.
(c) Breakdown of the active ingredient can be particularly serious with local formulations in which the carrier is not sufficiently "inert"; no scope exists for reapplication.
(d) Insect resistance depends on the insecticide used and the insect species, rather than on the formulation.

Among suitable chemicals are the established (malathion, lindane/gamma BHC) and the potentially better (pirimiphos-methyl, synthetic pyrethroids).

9.3.2 **Insecticidal sprays.** The method of using sprays is to allow 10-15 ppm active ingredient (ai), contained in the minimum of water necessary to give an even coverage (about 0.3-2 l/tonne, depending on the applicator). Such a small quantity of water will not cause moulding. The insecticide can be applied with a small domestic applicator (Shelltox-type), but a knapsack sprayer reduces labour requirements.

Application of sprays differs with the type of storage facility used. In warehouses, the following procedures are used.

(a) *Bagged produce.* Each layer of bags is sprayed as the stack is built; this should give protection for several months, but in case of reinfestation, the stack should be resprayed and fumigated for effective penetration.
(b) *Space spraying.* A non-persistent insecticide is sprayed to kill the adults of flying insects, especially warehouse moths; used in conjunction with fumigation under sheets.
(c) *Fogging.* For the same purpose, an electric applicator delivers very fine droplets, which hang in the air, maximizing effectiveness.
(d) *Surface treatment.* A persistent insecticide is sprayed on walls, roof and floor of the storage structure.

In cribs, the insecticide is sprayed directly onto the produce. If there is significant field infestation it is advantageous to spray each basketful during loading; otherwise, insecticides should be applied to the outside of the crib after loading, and reapplied at intervals as necessary (monthly is suggested).

Problems with sprays include the following:
(a) breakdown of chemicals in the highly ventilated crib environment (al-

though this implies a minimum residual toxicity for consumers);
(b) poor penetration with some types of structures; and
(c) lack of availability of sprayers and suitable chemicals.

Among suitable chemicals are the synthetic pyrethroids, usually used for space-spraying and/or control of the larger grain borer; dichlorvos for automatic fogging (**Note**: this is highly toxic to mammals); and malathion or pirimiphos-methyl for general use (since it costs less and has lower toxicity).

9.3.3 Fumigation.

Methods include fumigating produce in containers, or fumigating surfaces. The produce is placed in drums, plastic bags under tarpaulins or plastic sheets. After addition of the chemical, the produce must be kept in airtight conditions for at least three days for Phostoxin or about one day for EDB, depending on the doses applied. For fumigation of stacks in warehouses, it is necessary to spray the roof and walls simultaneously to prevent reinfestation. Grains must be protected from subsequent reinfestation.

Applications of fumigants differ with the type of crop they are used on. They are indispensable for export crops such as groundnuts, coffee and cocoa. At the small-farm level, fumigation might be justifiable for seed material of high-value crops such as grain legumes.

Fumigants can be very dangerous if incorrectly used; they should not be used in domestic living quarters. Another problem is that they have no residual action.

There are two types of formulation of chemicals used in fumigation.
(a) Phosphine gas (e.g. Phostoxin) is supplied as tablets of aluminium phosphide, which release phosphine on contact with moisture in the air. It is convenient to use, but requires airtight conditions for three to four days for total kill, and longer in cool conditions.
(b) With ethylene dibromide, methyl bromide, and carbon tetrachloride there are various combinations and formulations available (e.g. Trogocide). All are volatile liquid fumigants. Capsules and sachets are available for small-scale applications and pressure cylinders for large-scale ones. They are difficult to use and there is some residual toxicity and a possible consumer hazard: a shorter time is required for fumigation — normally less than one day — depending on the formulation used. Not recommended for farm- or village-level use, and only to be carried out by trained personnel.

9.4 Toxicity

All insecticides are also toxic to mammals to some extent.

The toxicity is usually expressed as an **LD_{50}**. Technically, this is the dose required in mg of active ingredient (ai) per kg of the body weight of the consumer, under specified conditions (method of application and time span), to kill 50 percent of the test population, which are usually rats.

The LD_{50} figure of a chemical is a reasonable indication of its toxicity to humans, and thus of the hazards involved in its use. It should be noted, however, that some compounds are highly active against particular sorts of animals; for example, some organophosphorus insecticides, such as fenthion, are very toxic to birds, including chickens.

For toxicity of recommended insecticides, see Section 9.6.

9.5 Formulations and dosages

Commercial insecticides consist of a quantity, usually small, of the toxic compound (the active ingredient) with other substances such as:
(a) inert "spreaders" such as talc, etc. in dusts and water in sprayers;
(b) surfactants, enabling the compound to be mixed with water and adhere to the pests and stored produce; and
(c) synergists, which may be included to increase the effectiveness of the active ingredient (e.g. piperonyl butoxide with pyrethroids).

The mixture is described as a formulation which comes in forms with particular characteristics:
(a) dusts, for dry application;
(b) wettable powder (wp) for mixing with water for spraying; and
(c) emulsifiable concentrate (ec) for spraying.

The concentration of the active ingredient in the formulation will always be stated, either directly, as in "malathion 5-percent dust" (i.e. 25-percent active ingredient); or indirectly, as in "Actellic 25 ec" (i.e. 25-percent active ingredient in solution).

Dosages may be expressed in three ways:
(a) the quantity of crude product (cp) (i.e. the solution from the bottle to be used), as in "40 ml in 5 l, to be applied on 1 tonne";
(b) a concentration for spraying, usually referring to the active ingredient concentration, as in "use a 10-percent solution"; or
(c) a concentration as a proportion of the quantity of the produce, referring to the active ingredient concentration, as in "apply dust at a rate of 10 ppm ai".

It is important to be able to convert from one basis to another. For example, the instruction "Actellic should be applied at 15 ppm ai" means that 15 g of active ingredient should be applied to every million g (or 1 tonne) of produce.
(a) Starting with a 5-percent dust, this means that 100 g of crude product (cp) contain only 5 g of active ingredient. 15 g are contained in 300 g of cp, so 300 g of dust per tonne (or 30 g to every 100-kg sack) are applied.
(b) Starting with 25 ec, this contains 25 ml ai per 10 ml cp. Taking 1 ml as weighing approximately 1 g, to obtain 15 g of ai, $\frac{15 \times 100}{25}$ (i.e. 60 ml) of solution are required.

The required amount of insecticide for spraying can be mixed in any convenient quantity of water; for example, in using Actellic 25 ec to spray a crib, either a Shelltox-type hand pump (in which case a 60-ml solution in 250 ml of water is required) or a cp3-knapsack sprayer (for which 60 ml might be mixed with 5 l of water) may be used. In both cases, the whole spray solution would be applied to 1 tonne of produce.

If a solution concentration is specified, calculations proceed similarly:
(a) to apply a 7.5-percent (ai) solution, starting with 25 ec formulation, 7.5 ml are required in every 100 ml of spray, equivalent to 75 ml in every l;
(b) there are 25 ml in each 100 ml of formulation;
(c) to obtain 75 ml ai, 300 ml of formulation are therefore needed;
(d) water is then added to make the total up to 1 l.

Note: 300 ml cp + 700 ml water provide 1 l of spray.

9.6 Some insecticides for use with stored products: summary of properties

Insecticides for use with stored products include the following.

Gamma BHC/lindane
- organochloride
- very stable
- acute oral LD_{50} for rats 88 mg/kg
- harmful to livestock and fish
- avoid use on food grains
- some insects are resistant to this compound

Malathion
- organophosphorus
- limited persistence
- LD_{50} 2 800 mg/kg
- general purpose, spray or dust

Iodofenphos (e.g. Nuvanol, Elocril)
- organophosphorus
- limited persistence
- LD_{50} 2 100 mg/kg
- general purpose, spray or dust

Synthetic pyrethroids (e.g. Permethrin)
- limited persistence
- LD_{50} 3 000 mg/kg
- sprays or dusts; expensive

Pirimiphos-methyl (e.g. Actellic)
- organophosphorus
- limited persistence
- LD_{50} 800 mg/kg
- general purpose, spray or dust

Dichlorvos (e.g. Nuvan)
- organophosphorus
- short-lived
- LD_{50} 80 mg/kg
- used for fogging in warehouses

Propoxur (e.g. Baygon)
- carbamate
- very persistent
- LD_{50} 100 mg/kg
- storage structures
- not very effective against beetles

10. Crib storage

10.1 Introduction

Traditional systems rely on natural ambient air for drying. Some drying takes place in the field before the crop is harvested. Producers of small quantities of commodities dry un-dehusked maize, for example, by exposing the cobs to the atmosphere. Larger producers use a variety of naturally ventilated structures to reduce the moisture content of the cobs to about 20 percent, when they are considered reasonably safe against insect attacks.

Round, slatted-wall structures are extensively used for drying in the humid tropics. They range from 1 to 3 m in diameter and are up to 2.5 m high. Platforms are used to store un-dehusked maize cobs, often arranged in a specific pattern, and a fire may be lit beneath the platform. A thatched roof provides protection from rain.

The main limitation in using traditional ventilated structures is the long period of field-drying required before the crop can be loaded into the crib. This leads to much higher levels of infestation at harvest time which may increase during storage.

Improved systems of maize harvesting, drying and storage should aim at the following:
(a) much earlier harvesting (when moisture content may be as high as 35-percent wb, reducing levels of field infestation);
(b) drying in improved traditional structures, with which farmers are familiar; and
(c) providing protection from infestation during storage.

10.2 Optimal crib design

A series of trials to measure the effect of the size and shape of cribs on the drying rates of dehusked maize cobs and on the levels of infestation and damage were carried out under the direction of the African Rural Storage Centre (ARSC) in Ibadan and Benin, Nigeria; these trials gave the following results:
(a) un-dehusked and dehusked maize cobs in cribs dried at the same rate;
(b) un-dehusked cobs with more than 26-percent mcwb became mouldy;
(c) orientation of rectangular cribs was relatively unimportant, except in areas with a prevailing wind;
(d) dehusked maize cobs required protection from insect damage, if they remained undisturbed in the cribs after drying;

(e) the drying rate of maize cobs in cribs, particularly narrow ones, was similar to the drying rate of cobs in the field;
(f) when initial moisture content exceeded 28-percent mcwb the 600-mm-wide crib gave the best results;
(g) roof overhang was important in reducing rain damage to the top surface layer of the stored cobs. Surface wetting of the sides or ends of the cobs was not important in most seasons;
(h) any wall material offering a minimum of 10-percent fairly evenly distributed open area was satisfactory;
(i) any material that sheds rain was suitable for roof covering;
(j) most effective insect control during storage was achieved by monthly spraying of the outside of the crib with a pirimiphos-methyl preparation.

10.3 Design of improved cribs

The recommended design is shown in Figures 5.4 and 10.1 and Section 5.6.2.

The base of the storage compartment is 1 m above ground level and is carried on 1.5-m vertical supports, separately from the longer vertical supports carrying the roof and walls of the crib. Rat guards are fitted to the vertical support below the crib floor level.

The width of the crib is 600 mm for very humid areas and up to 1 500 mm where drying conditions are better.

The capacity of the crib is given in the following table for each m in length.

Crib width (mm)	Weight of cobs at harvesting (kg)	Weight of grain (at 14% mcwb)
600	500	300
1 000	850	500
1 500	1 275	750

The crib may be made as long as required. Any materials used for its construction must support the weight of the cobs and provide a minimum of 10-percent open area.

10.4 Cost of crib construction

A crib made of sawn timber uprights, wire netting walls, corrugated-iron roof and metal rat guards will be three times more expensive than a similarly sized

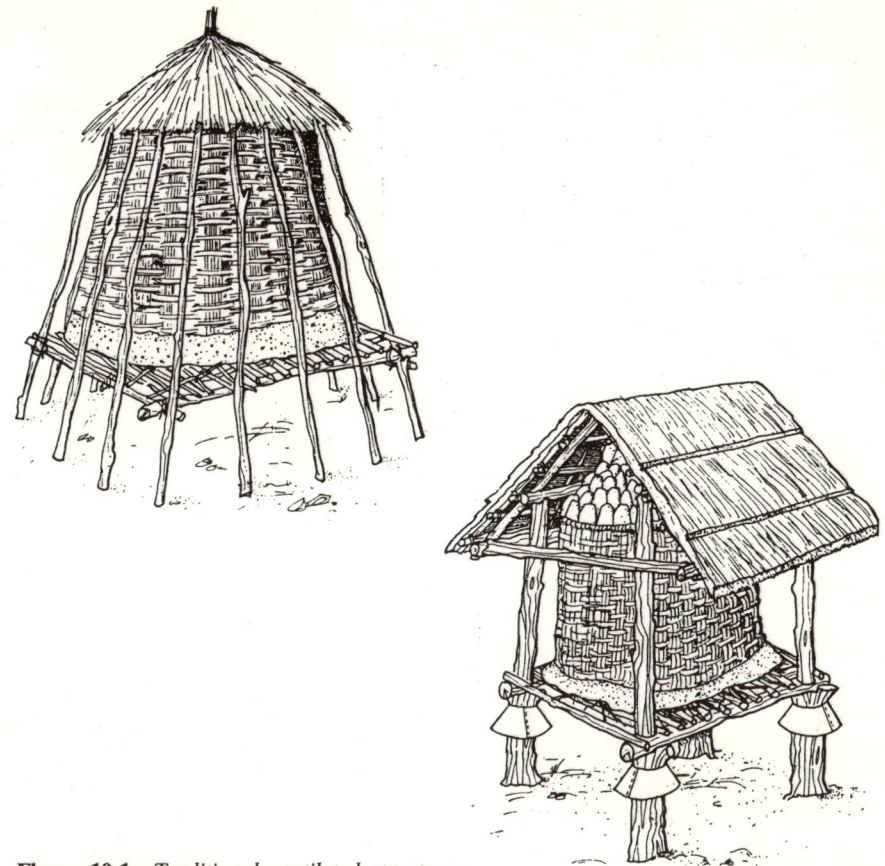

Figure 10.1 *Traditional ventilated structures*

crib using teak pole supports, bamboo slats for walling, a corrugated-iron roof and metal rat guards. A crib made entirely of home-grown materials from the farm or forest will be approximately half the cost of a crib made with teak poles, and one-sixth of that made with sawn timber.

These costs refer to capital cost per tonne stored per year, and take into account the expected life of the cribs.

Cribs constructed from home-grown materials require more maintenance than the more expensive constructions and even then may only remain serviceable for four seasons. Teak pole cribs may serve for eight to ten seasons if protected against termite attack.

Pamphlets describing the construction and use of an improved maize crib were produced by the Rural Structures Unit in the Ministry of Agriculture in Kenya and by the FAO/DANIDA African Rural Storage Centre, Ibadan, Nigeria.

11. Root and tuber storage

Harvesting is a most important function. Unless this operation is carried out with maximum efficiency, later PFL (prevention of food loss) activities may be a waste of time. If, for example, roots and tubers are bruised or otherwise damaged during harvesting, consideration of improved handling or packaging is not likely to be worthwhile, since an early infestation with moulds and viruses will occur and rotting will have started. If harvesting operations are correctly undertaken there is greater scope for later introduction of improved methods. Provision of the proper tools and equipment for harvesting and training workers in their correct use should be a priority PFL activity.

11.1 Yams

In most parts of West Africa, for example, the main crop of yams is harvested between September and November. Some will be consumed or sold immediately, but the bulk will be stored for a maximum period of six months for later use, either as food or for planting.

There are several traditional methods of yam storage:
(a) one of the most popular, particularly in the forest areas, is a bamboo enclosure (yam barn) containing a vertical framework onto which yams are tied individually with rope or bush twine. Shade is always provided, generally by trees;
(b) in the savannah zone less elaborate storages are generally found, such as shelters built of sorghum stalks with yams stacked inside;
(c) using clamps or underground storage has been successful.

Losses in these types of traditional storage are very high; a conservative estimate would be 25 percent. They are the result of a number of factors:
- the nematode, *Scutellonema*, commonly called the yam nematode, is one of the most important causes of loss in storage. Yam tubers infested with this pest at the time of harvest will not store. Another important nematode is the root-knot nematode, which, however, does not generally affect storability;
- fungal and bacterial rots are important. Many of these pathogens (wound pathogens) invade the tuber through wounds caused by nematodes or physical damage such as cuts and scrapes;
- rodents such as rats and mice, and occasionally other mammals, attack yams during storage, causing loss;
- insects may also contribute. Through their feeding, mealy-bug and scale

in particular reduce the food reserves in the tubers, often leaving them too weak to regrow when used as planting sets;
- physiological losses due to sprouting and respiration account for much of the weight loss in storage. These also deplete food reserves and exhaust planting materials. Sprouting often makes the yam bitter and unpalatable.

As a general rule, storage losses in the savannah zone are fewer than in the forest zone. This is probably because of the lower incidence of the *Scutellonema* nematode and to the tendency for yams grown in drier environments to have a higher dry-matter content. This appears to be important in storage, as is the fact that the cultivars grown in the savannah have a longer endogenous dormancy and do not sprout as early during the storage period.

Although it is recognized that storage is one of the critical problems limiting yam production, most discussions on the improvement of yam storage have failed to produce any new ideas in the area of intermediate technology. There appears to be no advantage in practice between the best traditional storage methods now used by the farmers and the large-scale advanced technology methods, which in any case would not be applicable in many parts of the rural world at present.

The following are important in reducing losses in traditional storage:
(a) it should be recognized that there are differences in storability between species of yams. As a general rule, water yam (*D. alata*) stores better than white yam (*D. rotundata*), and white yam better than yellow yam (*D. cayenensis*). Also, there are differences between varieties within a species. Varieties which are inherently better storers are those with (i) a long dormancy, (ii) good healing capacity and (iii) shapes easily dug from the soil with little injury. Poor storing varieties should be sold or consumed first and only varieties known to be good storers should be kept for long periods;
(b) the condition of the yam tubers when they are placed into storage is very important in determining storage life. Only sound yams will store; this means tubers which are free of nematodes, rots and physical damage. To reduce physical damage tubers should be handled carefully during all harvesting and transporting operations. Any injury, such as scraping, bruising or deep cuts from the cutlasses or hoes used for digging, will predispose the tuber to pathogen attack and subsequent rotting in storage. Tubers, especially those freshly harvested, should not be exposed for long periods to strong sunlight, as this may lead to injury and rotting;
(c) stored tubers should be checked routinely and tubers which are beginning to deteriorate should be marked for immediate consumption;
(d) if tubers are to be used only for food and not for planting, sprouting can be delayed by breaking off the head of the yam before tying in the barn or placing in storage. If planting sets begin to sprout before planting time it is not wise to break off the heads, but the sprout should be cut off as soon as they begin to elongate and be kept cut until planting time. This prevents the sprouts from exhausting the set before it is planted;

(e) storage areas, whether barns, cribs or pits, must be cleaned before any new yams are brought in. Old yams and debris should be removed from the store and surrounding areas. This avoids contamination by the pathogens carried in old decaying tubers. Piles of logs or pots, machinery, tools, etc. where rats and other rodents may hide should be cleared away. If trees are used for shading, they should be trimmed three months before storage begins so that sunlight can penetrate to the barn floor and sanitize it. Early trimming also enables the trees to regain the necessary degree of shading before the storage season begins. In yam barns insecticide dust may be sprinkled at the base of all vertical supports to control ants and other crawling insects, but great care must be taken to ensure that food yams never become contaminated with insecticide;

(f) it may be possible to modify the structure of traditional storages so that rat guards can be added.

Recent experiments indicate that the curing of yam tubers, using relatively high temperatures and humidities, can improve storage by healing wounds and toughening skins. Temperatures of 30-40°C and 70-90 percent humidity for one to four days are effective in reducing storage losses. These conditions can be achieved by various means; one of the easiest is to cover tubers with a tarpaulin.

In curing, timing is important. Curing should be undertaken immediately after harvest to heal the wounds inflicted during harvesting and subsequent transportation to the barn. Any handling after curing must be carried out with utmost care to avoid new injuries.

The advantages of curing are greatest when tubers are stored in reduced temperatures, and less when they are put into traditional barns or cribs. The reason is that traditional structures allow the yam tubers to cure during the early part of the storage period.

Two large-scale advanced technology improvements have been studied. Although they are not appropriate for storage on the farm, they may have a future application as part of the marketing system for bulk storage at collection points. These include the following methods:

(a) controlled temperature storages used throughout the world for a wide variety of perishable products may be used to prolong the life of yam tubers. Temperatures of around 20°C have proved effective in inhibiting sprouting and slowing down respiration. However, yams should not be stored below 15°C, or chilling damage will occur;

(b) a second method, the use of gamma irradiation, has also been tested. Dr Adesuyi of the Nigerian Stored Products Research Institute has shown that if tubers are irradiated before being placed into storage, sprouting is inhibited for up to six months.

These are methods of the future, but they are mentioned to indicate that research on yam storage is not being neglected.

12. Processing of cereals (other than rice)

The harvesting and threshing of cereals are just as important as the harvesting of roots and tubers. Unless the operations are carried out efficiently other PFL (prevention of food loss) activities may prove of little value. For example, the husks of grains should not be broken in harvesting; otherwise, insect attack and infestation will develop more quickly. Provision of the proper equipment for harvesting and threshing, and training in its correct use, are essential PFL activities.

The principal operations in cereal processing are:
- threshing
- grading
- milling
- sifting

12.1 Threshing

The process of threshing separates the kernels from the stalks or panicles on which they grow. Threshing may take place in the field, or at the homestead or village; it may be carried out manually with the aid of animals, or with machinery. A simple method consists of beating the cereal heads against a wall or the ground; animals or humans can also trample the panicles on a hard surface, or animals can draw a machine or sledge over the grain.

Threshing machines may be powered by humans or animals or, in more sophisticated forms, by internal combustion engines. Many designs have been field-tested and found to operate satisfactorily.

Maize grains must be separated from the cob after the husk has been removed. A variety of manual and powered systems are available for this operation.

12.2 Grading

Grading consists of separating the sound kernels from chaff and impurities, and may be achieved by sieving or winnowing.

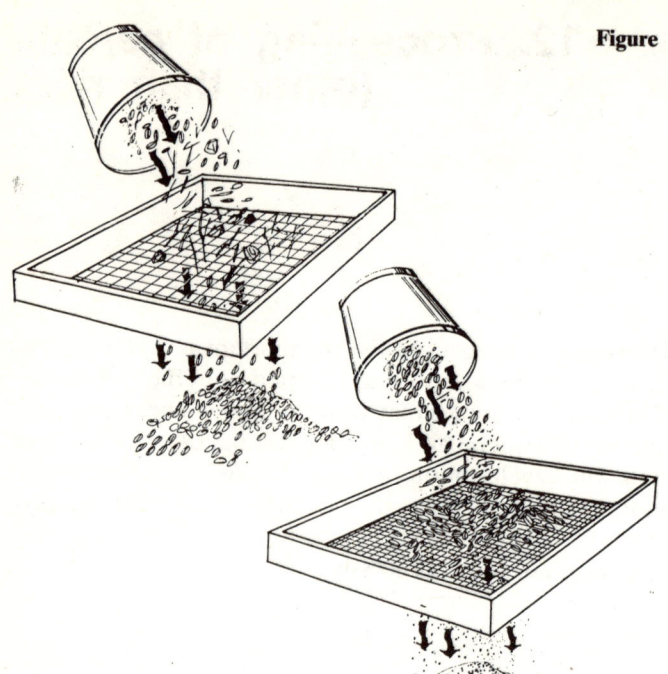

Figure 12.1 *Manual sieving*

12.2.1 **Sieving.** Impurities are separated on the basis of their differences in size from the kernels. Hand sieves are usually used singly. The simpler machines will have two sieves: one with oversized holes (which retain large impurities and let the grain kernel pass through), and one with undersized holes (which retain the kernels but allow smaller impurities to pass through).

12.2.2 **Winnowing.** In this process impurities are separated on the principle that their density differs from that of the grain kernels. The operation depends on air movement to remove the lighter fractions. The simplest method is to drop a basket of kernels and impurities in a thin stream onto a clean surface through a slight natural breeze. This is a slow and laborious process but it is still widely practised.

Winnowing machines operate on the same principle, but air movement is created by a fan.

12.2.3 **Selective picking.** An internally indented cylinder will remove impurities which are smaller than the grain kernels by carrying them beyond the point at which the sound kernels are ejected and into a trough as the cylinder is rotated.

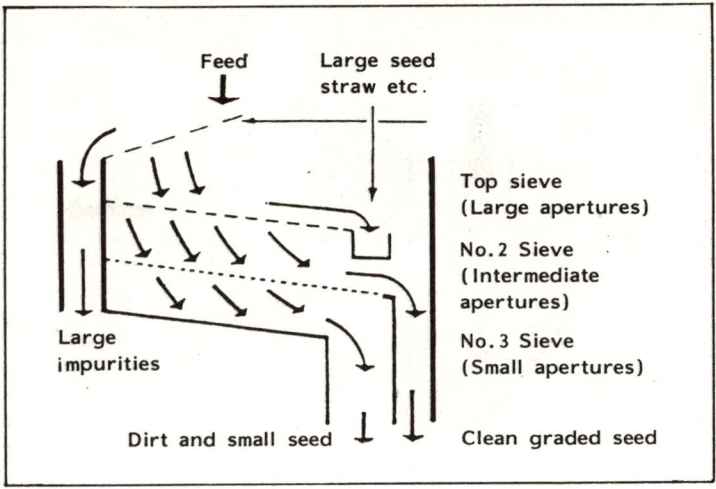

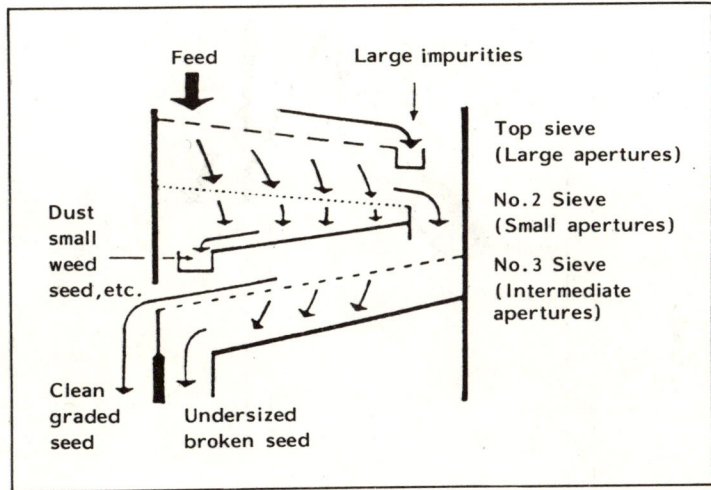

Figure 12.2 *Three-sieve separator*

Sophisticated machinery (SORTEX) is available for sorting individual kernels according to their colour, but these are expensive and are used only in specialized applications.

Hand-picking is an effective but tedious operation that is nevertheless widely practised by farmers.

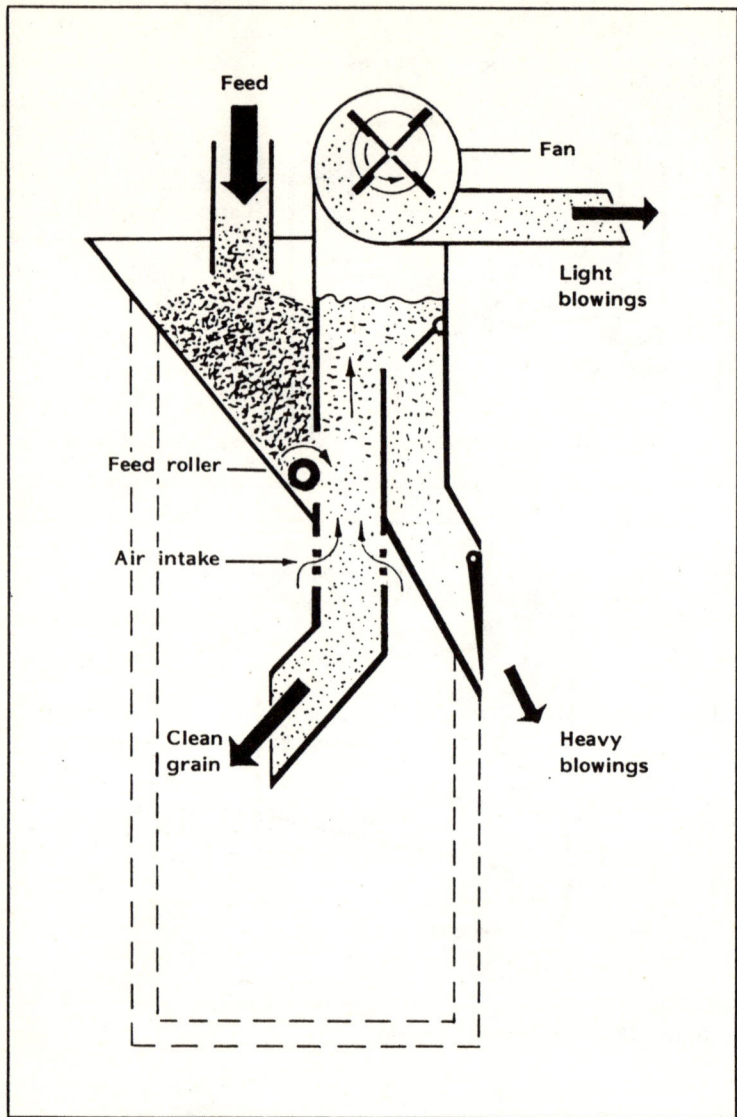

Figure 12.3 *Simple aspirator*

12.3 Milling

Milling involves the production of flour from the endosperm of cereal grains. In most cereals, including maize, the seed coat is first removed by stripping (either by hand-pounding after soaking or in a hulling mill) before being milled to make flour. Milling may be by hand-pounding in a mortar, by

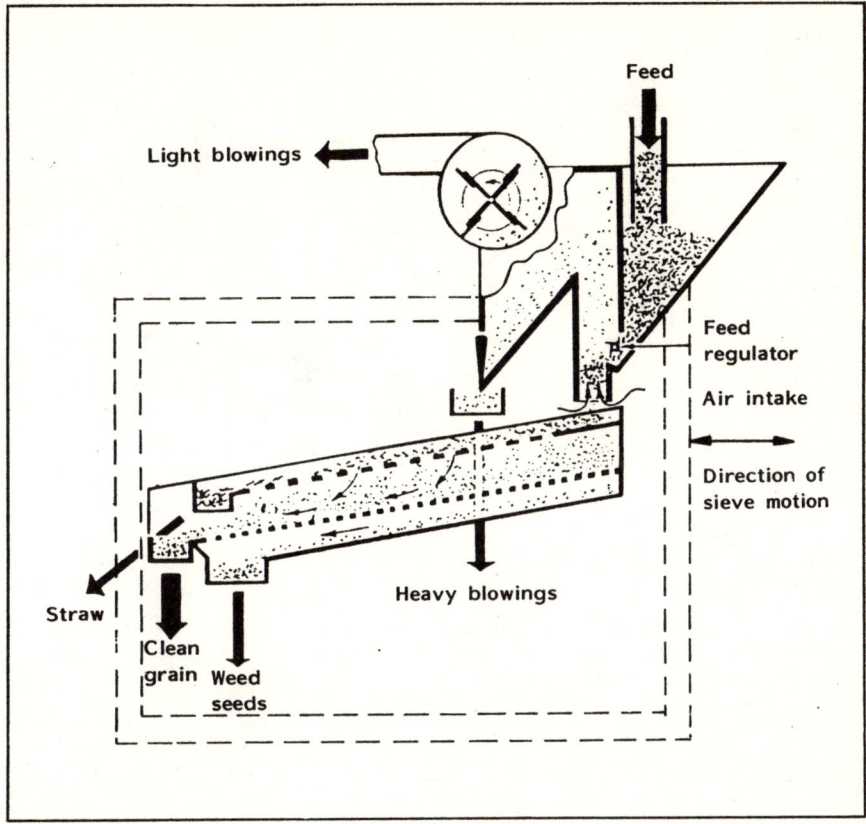

Figure 12.4 *Two-sieve cleaner with head aspirator*

forcing the grain between two stones, or by using mechanically powered hammer, plate or roller mills.

12.3.1 **Milling equipment.** The most widely used mills at village level are the plate mill and the hammer mill; for commercial operations, the roller mill is most common.

The plate mill consists of two circular cast-iron plates with surface burrs mounted on the same horizontal axis, so that the plates are vertical. One plate is held stationary and is attached to the frame of the mill; the other is mounted on a driving shaft and can be adjusted to alter the clearance or separation from the fixed plate. In operation, grain is introduced through the centre of the stationary plate and is ground when it passes between the plates to their edge. Here the flour is collected and discharged from the outlet spout. Some models have three plates, the two outer ones stationary and the central one rotating.

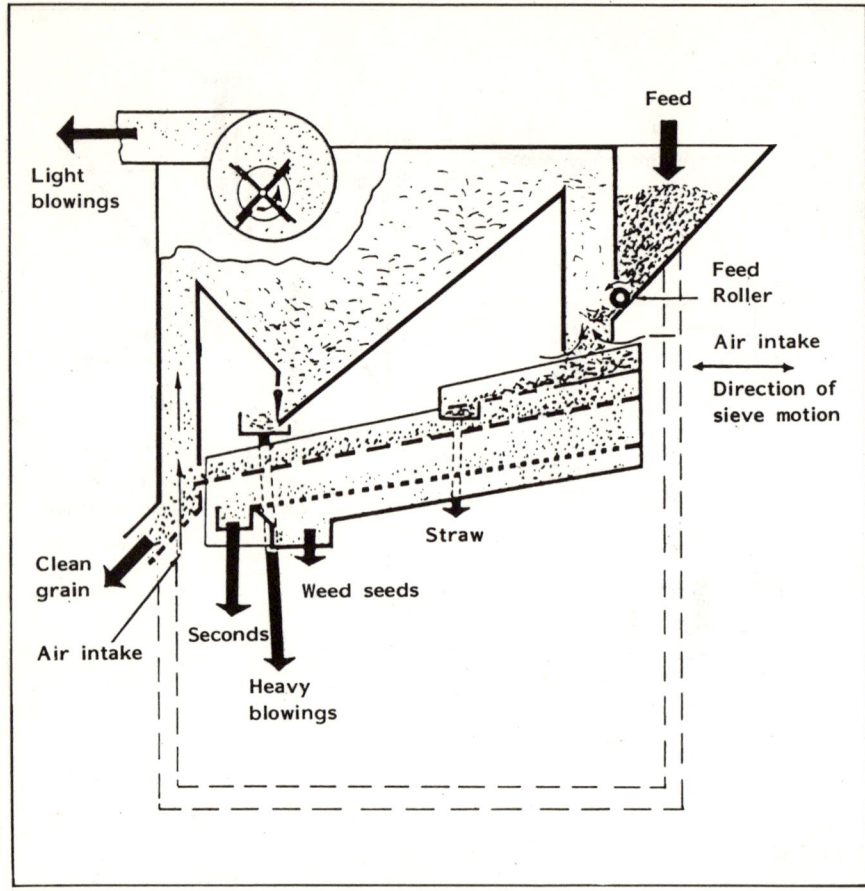

Figure 12.5 *Three-sieve cleaner with head and tail aspirator*

Great care is needed to ensure that the plates do not make contact. They must be separated, using the hand-wheel, when the mill is running empty. A feeding device is necessary to provide an even flow of material to the intake point, which is usually located at the centre of the fixed plate. The plate separation is adjusted by a hand-wheel which moves the driving shaft and its rotating plate in its plain bearings against a compression spring.

The shaft rotates at low speed — usually less than 1 500 rpm — and the output of a typical mill with a 5-hp motor and 270-mm diameter plates is approximately 250 kg/h. The plate mill will grind wet cereals, which would block a hammer mill. The fineness of grinding in a plate mill depends on the following:

Figure 12.6 *Winnowing*

- type of plate and burr pattern being used
- speed of rotation
- condition of the plate surfaces
- pressure on the plates
- feeding rate
- type of grain
- moisture content of the grain

Small diameter plate mills may be hand-powered.

The original hammer mill was developed from the hand-pounding system. The hand-held pestle was replaced by a heavier wooden hammer fixed at the end of a lever pivoted near its centre. When at rest the hammer rested in a hollow (often cut from rock) into which grain was placed. Pounding was achieved by pressing down on the other end of the lever, which raised the hammer, and then releasing it to fall under its own weight.

Modern hammer mills consist of a set of fixed or swinging hammers mounted on a rotating shaft and surrounded by a perforated metal screen. The shaft is rotated at up to 6 000 rpm, depending on the design and diameter of the hammers, which usually have a speed of 75-100 m/s measured at the tip of the hammers. The grain is introduced into the path of the rotating hammers through a slot in the screen and the ground material is discharged through the screen.

A paddle-bladed fan draws air through the screen and delivers the flour to the outlet spout, where it is separated from the air flow by a cyclone and/or woven-textile filter bags.

The modern hammer mill is excellent for the fine grinding of dry cereals. It is not damaged by running empty and can be powered easily from high-speed internal combustion engines or electric motors. Output is generally about 60 kg/h per kW of power input. Up to 25 percent of the power is consumed by the fan, which not only removes the ground meal but provides the necessary flow of air through the screen.

The fineness of grinding depends almost entirely on the size and shape of holes in the perforated screens which partially or completely surround the hammers. The grinding action results from friction of the grains being repeatedly forced against the screen and against each other and, particularly with brittle materials, from impact with the hammers. The hammers are usually reversible to compensate for wear.

The roller mill is a more sophisticated form of mill than the plate or hammer mill and is used for producing high-quality fine flour, generally from wheat but also from maize and sorghum.

The precision-cast steel rolls have fluted surfaces and rotate in opposite directions at slightly different speeds. The clearance between the rolls can be set precisely, so that when fed with a single layer of carefully graded grain a small predetermined amount from each grain's surface is removed as it passes vertically downward between the rolls. The whole grinding operation consists of passing the grain through a series of such mills in succession, in up to ten possible stages. The output from each stage is sifted and the operation allows the various constituent parts of grain, such as germ and bran, to be collected separately. These mills have a high output and usually produce flour for the urban population.

13. Small-scale rice milling

13.1 Introduction

The production of white rice from paddy is complex and involves many operations. In large-scale plants the machinery and equipment used are very specialized, with each item only carrying out perhaps a single operation of the 20 or more that may be required for commercial rice milling. Large-scale plants must operate at high capacity to justify the investment in equipment.

In small-scale rice milling, with capacities up to 500 kg/h, a piece of machinery will carry out several of the operations in producing white rice from paddy, either in a single pass through the machine, or in several passes, with machine adjustments being made between each pass. Two or more identical machines may be used in successive stages of the process, each being adjusted to perform a specified task.

13.2 Stages in rice processing

The various stages in rice processing are shown in Figure 13.1 covering the operations from harvest of the panicle to the production of graded, polished white rice.

The moisture content of harvested paddy will usually be in excess of 20-percent mcwb. This must be reduced to 12- to 14-percent mcwb for efficient hulling and processing operations. Paddy can be hulled outside this moisture content range but the performance of the machines is poor. The normal pre-hulling operations are as follows:
(a) *parboiling*. This is a process which involves soaking the paddy, then steaming and drying it. Parboiling improves the nutritional quality of the rice, makes the hulling operation much easier, and gives a greater proportion of whole-grain white rice. Parboiled paddy must be dried before milling. Rice milled from parboiled paddy stores better than non-parboiled rice, and has a different taste, colour and cooking properties. Parboiling is a costly operation but its benefits generally outweigh its cost;
(b) *there are two main methods of drying*. The prevailing local method is sun-drying. The paddy is spread out on a clean surface (tarpaulin, concrete slab or even smooth, clean earth) and regularly turned by hand. Excessively rapid drying results in the development of hairline cracks in the endosperm of the paddy grain (sun-checking). These cracks enlarge and produce a higher pro-

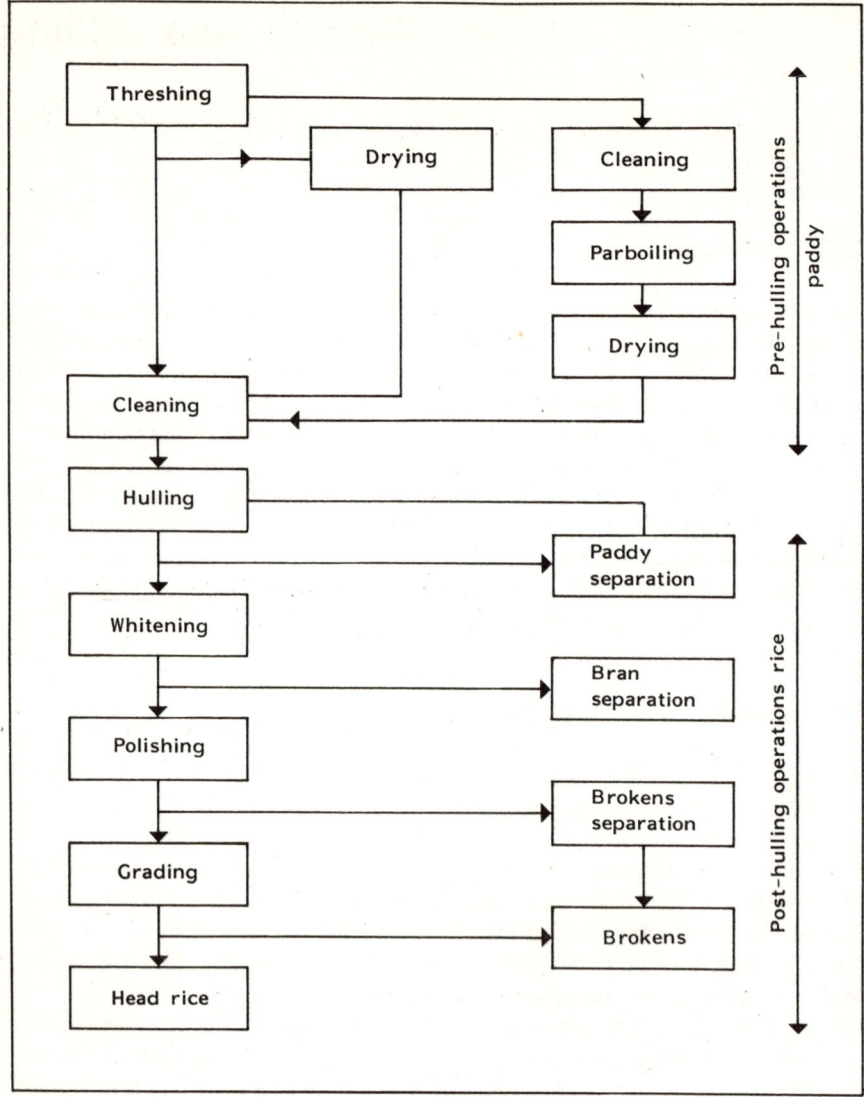

Figure 13.1 *Stages in rice processing*

portion of broken grains during subsequent operations. The incidence of cracks is reduced by a slower rate of drying which, in turn, can be achieved by increasing the thickness of the layer of paddy during sun-drying up to 150 mm, and by frequent stirring.

If artificial drying is employed the manufacturer's instructions should be followed. With very wet paddy, and particularly after parboiling, it is common practice to dry in two stages separated by a resting period during which the paddy is aerated.

Cleaning is an important operation; small stones and pieces of metal can damage the huller, while pieces of straw may cause an uneven flow of paddy to the huller. All impurities should be removed before the paddy is hulled. A combination of sieving and aspiration is commonly employed to separate the light impurities and a de-stoner is used to remove denser impurities.

If the paddy is to be parboiled before hulling, it should be washed and drained before being soaked, in order to remove soluble impurities which may otherwise discolour the grains.

13.2.1 **Hulling operations.** During this operation the hull (or husk, or shell) is removed from the paddy grain to produce brown rice. The husks have

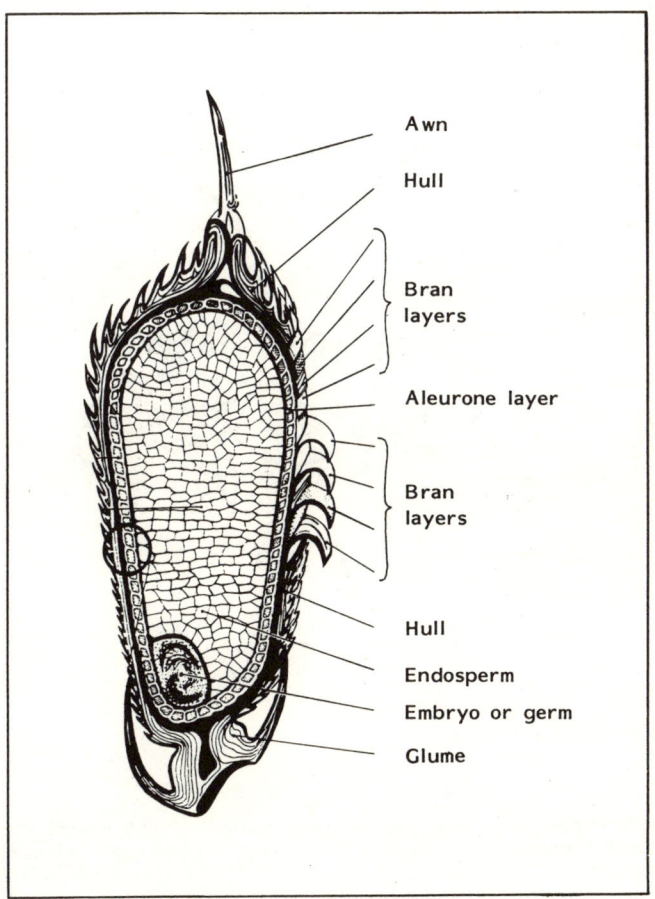

Figure 13.2 *Diagrammatic enlarged vertical section of a paddy grain*

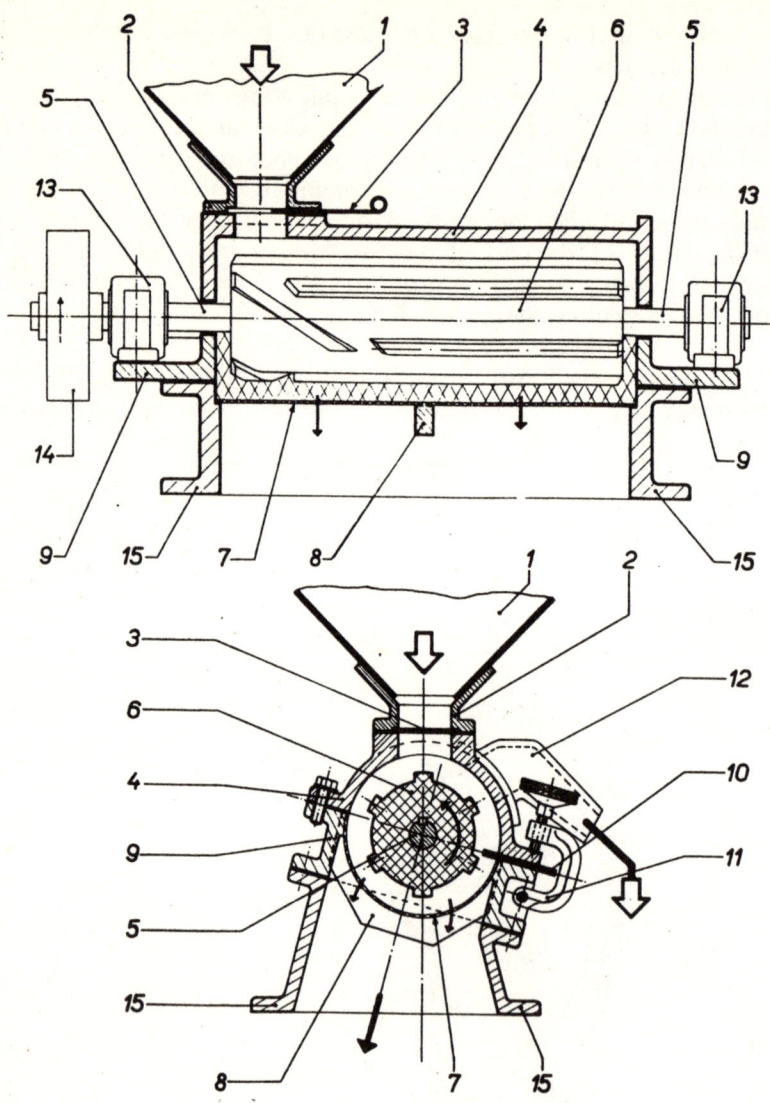

Used for paddy hulling and whitening

Key

1. Feed hopper; 2. Hopper base; 3. Feed regulating gate; 4. Cover; 5. Shaft; 6. Steel roll/shaft; 7. Screen; 8. Screen holder; 9. Frame; 10. Hulling blade; 11. Cover clamp; 12. Outlet spout; 13. Bearings; 14. Drive pulley; 15. Frame.

Figure 13.3 *Engleberg-type steel roll huller*

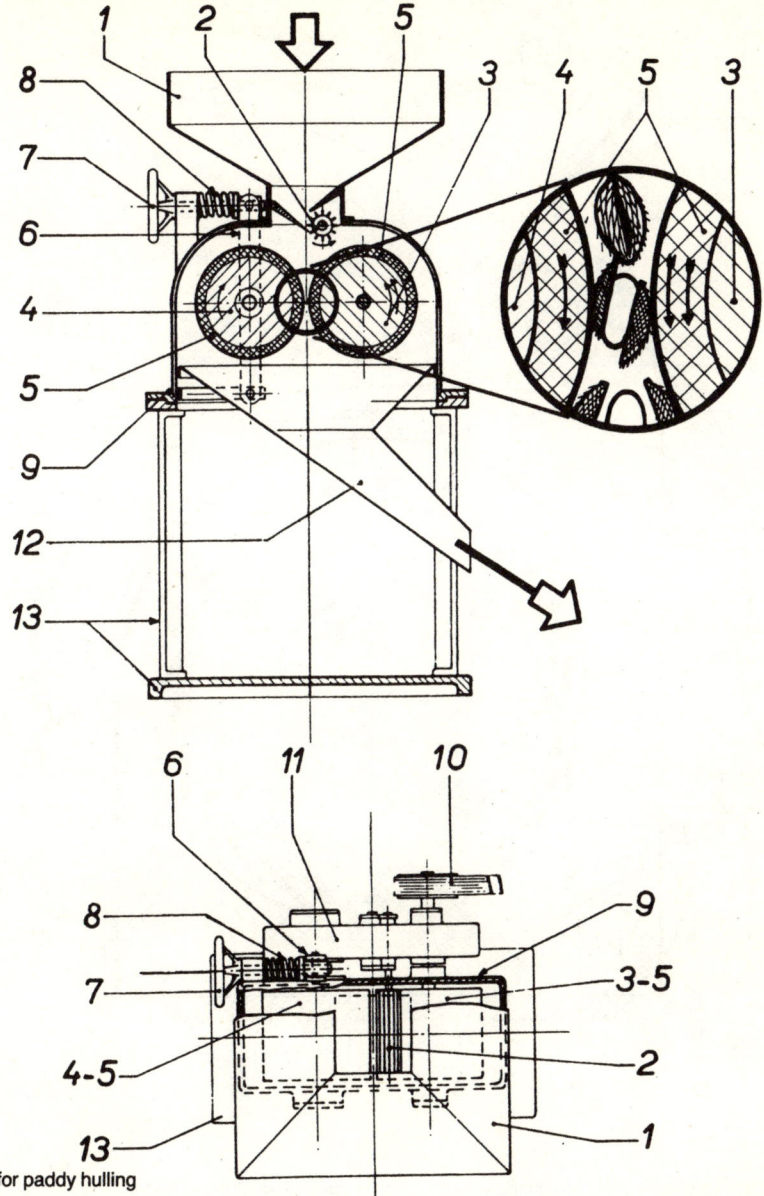

Used for paddy hulling

Key

1. Feed hopper; 2. Feed roller; 3. Fast roll; 4. Slow roll; 5. Rubber surface; 6. Roll adjusting arm; 7. Roll adjusting hand-wheel; 8. Compression spring; 9. Housing; 10. Drive pulley; 11. Drive unit housing; 12. Outlet spout; 13. Base and frame.

Inset: hulls being removed between rolls.

Figure 13.4 *Rubber roll huller*

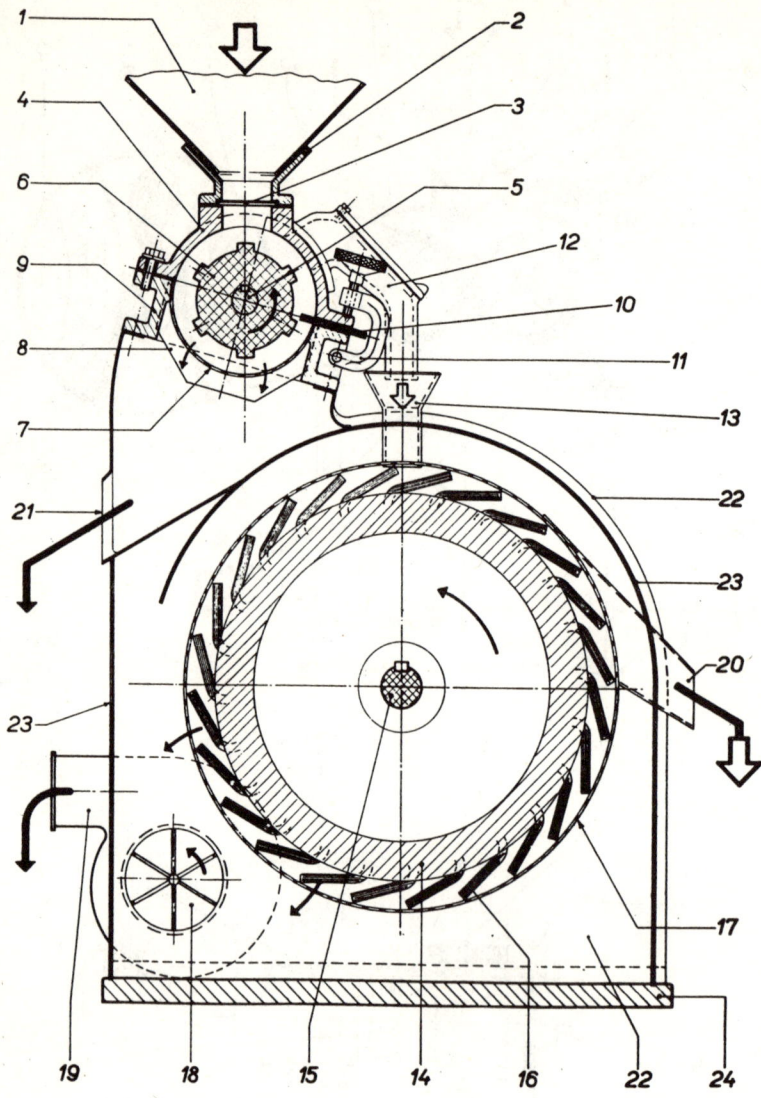

Used for paddy hulling and polishing, whitening brown rice and polishing white rice

Key

1. Feed hopper; 2. Hopper base; 3. Feed regulating gate; 4. Cover; 5. Shaft; 6. Steel roll/shaft; 7. Screen; 8. Screen holder; 9,22,23,24. Machine frame and cover; 10. Hulling blade; 11. Cover clamp; 12. Outlet spout; 13. Grain inlet to polisher; 14. Drum; 15. Shaft; 16. Leather strip; 17. Screen; 18. Fan; 19. Fan outlet; 20. Grain outlet; 21. Bran and husk outlet.

Figure 13.5 *Engleberg-type steel roll huller and polisher*

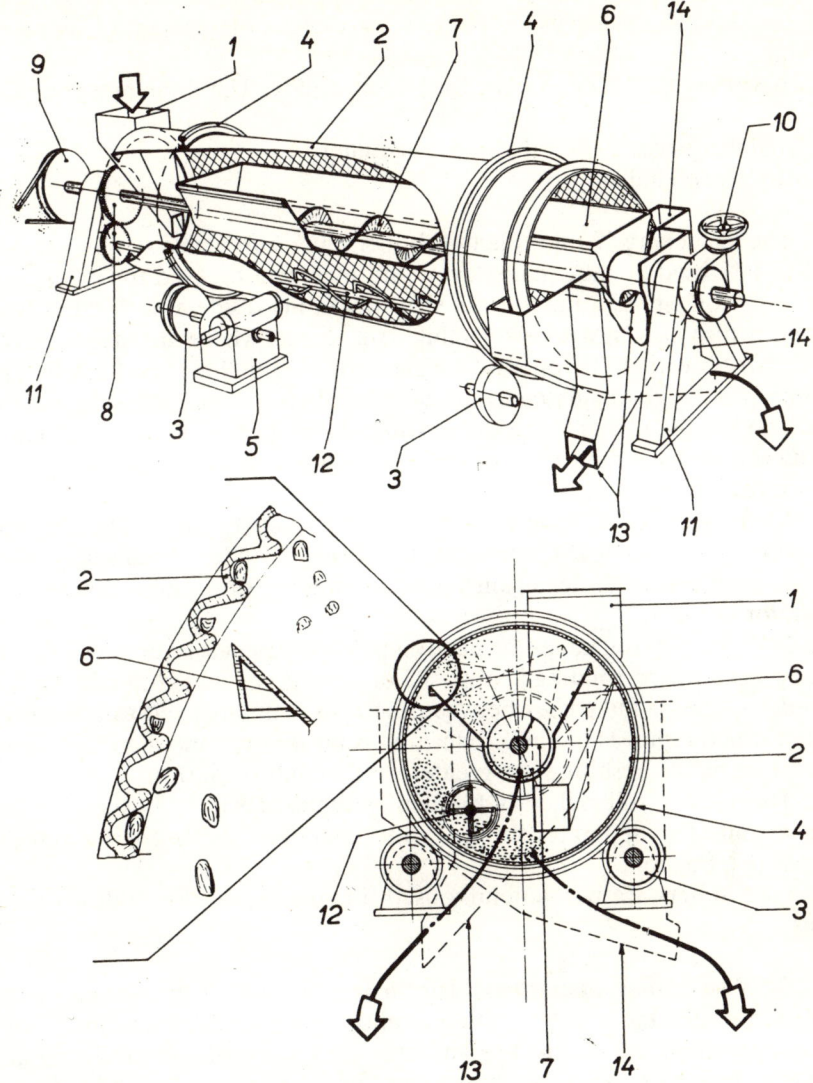

Used for paddy grading and cleaning, hulled rice grading, and rice grading

Key

1. Feed hopper; 2. Indented cylinder; 3. Cylinder supporting roll; 4. Outer cylinder ring; 5. Speed reducing gearbox; 6. Collecting tray; 7. Screw conveyor; 8. Conveyor driving gears; 9. Drive pulley; 10. Hand adjustment for tray position; 11. Frame; 12. Grain spreader; 13. Liftings outlet; 14. Grain outlet.

Figure 13.6 *Indented cylinder cleaner*

no nutritional value but may be used as fuel, perhaps in the parboiling operation (see Fig. 13.2). The ash can be used as a source of pure carbon for steelmaking.

A variety of machines exist for hulling paddy. The three most popular are:
- Engleberg-type steel roller and cage huller
- Rubber roll huller
- Disc huller

The Engleberg (Grant, Planter) huller represents an old design which is still widely used at the village level, and may also be used for processing maize. It consists of a fluted steel shaft operating inside a perforated steel screen, which also carries a projecting strip of steel whose distance from the shaft can be varied. To operate satisfactorily the huller must be full. The degree of hulling is regulated by the clearance between the steel strip and the shaft, and by the rate at which the mixture of rice, husks and unhulled paddy is allowed to discharge from the working chamber. An adjustable slide controls the discharge rate.

Various accessories may be added to the Engleberg huller. The common attachments are a polisher, consisting of rotating cylinders fitted with leather strips which press the rice against a perforated housing, and a simple husk aspirator.

In the rubber roll huller the paddy is passed in a single layer between rubber-covered rollers rotating in opposite directions and with different surface speeds; as the paddy passes between the rolls it is subjected to a shearing action which removes the hull. Its action is far more gentle than that of the steel shaft huller, resulting in a greater yield of unbroken rice.

The rubber roll huller is frequently supplied with a husk-aspirating attachment. This separates the hulls and immature paddy grains from the brown-rice fraction.

Disc hullers are not generally used for small-scale rice milling operations.

13.2.2 **Post-hulling operations.** The main post-hulling processing operations are whitening, polishing and grading. In large-scale processing plants, these operations are multi-stage and utilize a succession of specialized machinery. In small-scale processing some operations may be omitted (e.g. grading); they may be unnecessary (e.g. the steel roll mill also removes the bran layers), or may be carried out by a second hulling machine that is appropriately adjusted.

Whitening refers to the removal of the bran layers as a separate operation after hulling. These layers closely adhere to the endosperm and have to be removed by rubbing against an abrasive surface and against other grains. The Engleberg-type huller can remove husk and bran in one operation.

Polishing is the final, more gentle stage consisting of cleaning bran particles and dust from the white rice and smoothing its surface so that it looks better.

Whitening and polishing operations are combined in the steel roll mill with polisher attached.
sary if the white rice is to be sold, or stored for more than a few days. Broken grains deteriorate more rapidly than whole grains and whole kernels usually command a higher price.

Grading is carried out by sorting machines on the basis of grain size (using sieves or an indented cylinder cleaner), grain density (using aspiration) or a combination thereof.

14. Sociological, economic and institutional implications of the prevention of post-harvest food losses

Physical losses occur at various stages after a crop has matured and before the food is consumed. Losses may be reduced at any stage of the post-harvest system by improved harvesting, drying, storage, processing or handling methods. The processes and operations, however, are interrelated and are subject to climatological, sociological, economic, agronomic, cultural and ecological conditions imposed by the environment in which they take place.

The effectiveness of any action undertaken to reduce losses must be economically justifiable, and also practical within the prevailing post-harvest system. In attempting to reduce or even assess post-harvest losses, it is essential that the functioning of the system in any particular environment be fully understood and analysed. Only then may the constraints, problems and eventual solutions or improvements be identified.

In Sierra Leone, for example, parboiling of rice is practised because such rice has a higher nutritional content for which consumers are willing to pay more. The milling of parboiled rice is also easier, as despite poor milling equipment fewer broken grains are produced and less loss incurred. Whereas in Sierra Leone the provision of parboiling equipment is appreciated, in Malaysia, parboiled rice is regarded as a food for the lower classes and many Malaysians are willing to pay extra for well-milled white raw rice with a low percentage of broken grains. Hence, extending the practice of parboiling here would not be generally acceptable.

Earlier sections of this manual have dealt with loss assessment, as well as the main technological and biological aspects of the post-harvest system. As already noted, the system must also be considered in its entirety before the adoption of innovations. It is therefore necessary to take into account such factors as cost-effectiveness, the institutional framework (including the marketing system), labour availability and consumer preferences.

14.1 Economic justification

Post-harvest treatment of any commodity is only undertaken where it will result in a reward to the owner. In a subsistence economy the activity may be the storage of grain or tubers, where the benefit arises from the longer period over which consumption of the commodity may be enjoyed. Harvests usually occur at the same time, leading to a glut of produce that cannot be consumed immediately. Part must therefore be stored if it is not to be lost.

In a mixed subsistence and cash economy or where a crop is produced solely for cash sale, producers will only make those changes in post-harvest procedures which they consider will increase their revenue. It is hoped that PFL (prevention of food loss) activities will lead to increased cash returns, but they will only be adopted if the cost/benefit relationship of the operation is favourable, and when the markets are able to absorb the increased supplies at a price which is profitable to the producer.

The cost of PFL activities depends on many factors. PFL project activities are normally concerned with the introduction of techniques to reduce physical losses and improve the incomes of small-scale farmers. They are concerned with improved handling, storage and primary processing of grains, pulses and roots and tubers, and with techniques to maintain the quality of fruits and vegetables. Activities have included providing farm and village storage structures, designing and constructing warehouses, providing small-scale driers, improving processing facilities (from threshing equipment for rice to field-grading and packing of fruit and vegetables), improving rodent and insect control measures and undertaking training activities on all aspects of loss reduction during the post-harvest period.

It is important that the initial cost-benefit analysis be positive. According to some reports a cost-benefit ratio of 1:1.5 is insufficient to persuade individual farmers to take the risk of introducing a change in a post-harvest activity, but a ratio of 1:2 is likely to provide the necessary incentive. This may be taken as an important guideline by both planners of activities to reduce post-harvest losses and those who have the responsibility for project implementation and for training in these matters.

For example, the provision of bin storage made of sheet metal to a farm or village would undoubtedly reduce losses in grain, but the initial cost could be so high in relation to the extra grain saved in the short run that other farmers would not be interested. On the other hand, where the cost is small a development will be repeated, such as using a straw and mud-coated container with a little malathion insecticide. In this case, only the malathion has to be purchased, while the straw and mud can be gathered and the container made with family labour. In the Scarcies area of Sierra Leone, for example, rice is stored in large wooden boxes, which many houses in the area have. The boxes are about $2 \times 1.5 \times 1.5$ m and are made from planks of hardwood which were once readily available in that country. They are impervious to rodents and are often an integral part of the house. The initial cost was negligible and the boxes have endured for many years. This specific example is given to demonstrate the utility of using locally available, inexpensive materials and methods.

In a cost-benefit assessment, another factor to take into account is whether the commodity is for home consumption or for cash sale. If an improvement of quality only is related to home consumption, producers will be reluctant to pay cash for the innovation. The introduction of simple crop driers has been of interest where the crop is consumed by the family, although discolouring and off-flavours may in fact develop. The situation is different when the crop is for cash sale, particularly if the sale prices are appreciably different for varying levels of moisture content or admixture content. Normally, the producer will wish to take steps to reduce imperfections in order to attract the highest prices, but price differentials may not be sufficiently wide to provide an incentive for the producer to improve quality. Those responsible for price-fixing policy in grain procurement authorities should note that a sufficient price incentive for well-dried grain (usually 14-percent maximum moisture) will remove the burden of drying from the authority and encourage efficient drying on the farm. This ensures that drying is done more quickly, thus promoting a viable PFL activity for the producer, reducing physical losses, and, at the same time, greatly reducing the authority's operating costs.

An important consideration in the cost-benefit ratio is provision for replacement of the capital assets. Tools, machinery or storage provided under such a PFL activity need repair and maintenance and will eventually wear out and require replacement. These factors should be taken into account in the initial costing estimates for the activity.

In analysing cost-benefit ratios, it is important to be as accurate as possible. It is easier to arrive at costs than to quantify benefits. Costs may arise which were not anticipated, and they should therefore be estimated on the higher side. Benefits will usually be based on estimates of future selling prices unless sales are being made to an agency, such as a marketing board, that has already declared its buying price for the next season.

The following example shows how a cost-benefit analysis should be made. The example is hypothetical, but the costs and prices are based on actual figures recorded in Indonesia in 1983.

In a village producing cassava in Southeast Asia, it is proposed to introduce a solar drier as a PFL activity. The product, cassava chips, is destined for sale to a processing factory which will pelletize the chips for export. One problem facing the processing factory is that normal sun-dried cassava chips are contaminated with dirt and fungal growths, are discoloured and may have acquired off-flavours. Animals and people walk over the drying material, there are showers of rain, and the wind blows dust onto the drying chips. For this product the processing factory will pay Rp 40/kg. For clean, uncontaminated cassava chips, the factory will pay Rp 45/kg. Use of a solar drier prevents this contamination. All the materials for constructing a solar drier are available in the locality or may be purchased in the village or the local "kecematan" (county) town. The capacity of the drier is 1 tonne of cassava chips, and the drying time is three days. Cassava production and processing is a year-round activity, which means that about 120 tonnes of chips could be dried annually. Assuming that only 50 tonnes will be dried in the year, the completed drier should last for several years, subject to repair and maintenance; but the initial cost will be covered within one year. It is also assumed that the piece of ground in the village on which the drier is built will be provided free, since many villagers will use the facility. Also, producers will provide their own labour to fill and empty the drier, as they do in spreading chips on the floor in the traditional way. Other costs such as handling, packing, sacks and transport are the same, whether drying traditionally or using the solar drier.

Cost of construction of the solar drier (1983, Rp 970 = US$ 1)

48 bamboo poles each 6 m long	Rp	250 each Rp	12 000
110 m white plastic sheeting, 0.10 mm thick		200 m	22 000
2.5 m black plastic sheeting, 0.10 mm thick		200 m	500
1 kg nails		200 kg	200
1 roll yarn		200 each	200
9 pieces wood board 2 m × 25 cm		1 000 each	9 000
8 pieces wire netting, No. 18 gauge 2 m wide		300 each	2 400
1.5 q charcoal		10 kg	1 500
4 l tar		500 tin	500
Labour, 7 days (1 carpenter, 2 labourers)		5 000 day	35 000
		Rp	83 300
	or, approximately Rp		90 000

Benefits

Normal return for 50 tonnes of cassava chips at Rp 40/kg	Rp 2 000 000
50 tonnes of clean cassava chips at Rp 45/kg	2 250 000
Additional return from using solar drier	Rp 250 000

Cost-benefit ratio: 90 000:250 000
That is: 1:2.7
This should therefore be an attractive proposition.

14.2   Institutional factors

If a PFL activity is successful it means that larger quantities of a commodity are available for sale. In some cases a surplus will be created for the first time, in others it will be an enlargement of the marketable amount. This could put pressure on the marketing system. Larger quantities may have to be stored, transported and sold. It is therefore important that the marketing chain operators, whether private traders, cooperatives or government organizations, be made aware of the increased supply, so that it may be absorbed. This consideration grows in importance the more remote the siting of the PFL activity is from the recipient market. Again, the cost factor in absorbing the increased produce from the prevention of losses must be taken into account in the cost-benefit assessment at the planning stage.

Introduction of the grading and packing of fruit and vegetables for the export market leads to reduced losses during transport and marketing operations, and increased income will be earned. Consumers of fruit and vegetables in local urban markets, however, may have low income and require cheap supplies. A PFL activity to reduce physical losses in such a situation may eventually prove to be a bad idea, as the increased costs incurred in grading and packing would be reflected in the sales price and might reduce sales. If so, the increased return per unit of sales may be less than the cost of grading and better packing. Therefore, the idea may not be practical, depending largely upon the market-consumer sector concerned.

This example does not necessarily imply that grading and packing are not recommended for internal markets. A limited operation may well be justified, but any proposal should be subject to a cost-benefit assessment. Initially, a pilot-scale operation should be undertaken with certain selected farmers to test the market. Where spoiled fruits or vegetables are discarded, the value of the satisfactory produce remaining may be enhanced; and, again, packaging made of cheap local materials will certainly reduce losses during handling and transport. Packaging may also facilitate retail display, further reducing losses. It is not possible to lay down firm guidelines in these matters, since the location, nature of the market, cost and availability of packaging materials, tradition and consumer acceptance must all be taken into account. Only local trial shipments on an experimental basis can determine the feasibility of an improvement idea. If, after the trial shipments, a positive benefit can be shown from an innovation, then there is a strong case for believing that the new development will become common practice and be repeated elsewhere. In the Indonesian example given in the previous section, the appeal of the solar drier would have been much less convincing to other producers if the processing factory price had fallen later in the year, although the same differential between normal and clean prices may have been retained.

Marketing systems vary from country to country with differing degrees of governmental involvement.

The marketing system has a bearing on PFL technology and may influence the extent to which improved practices are economic. In state-planned economies this means planned production, planned supply and planned pricing, but this has often resulted in high losses in fruit and vegetables because incentives to avoid losses at the various stages of the production/marketing chain were lacking. Under a government-operated marketing system, the main concern is often with quantity (the number of tonnes of fruits and vegetables distributed), with little emphasis put on quality (the quality of the produce distributed and the cost of the task carried out). In this situation, technical knowledge of post-harvest measures alone would not contribute to the reduction of losses.

Whatever the marketing system — state-controlled, free enterprise or otherwise — it must be effective. If not, it will mean high prices for the consumer and/or lower returns to the farmer. In both cases losses will be high. Marketing agencies carry out the marketing function; these can be private individuals (such as farmers who may be wholesalers and money-lenders), or companies, cooperatives or government corporations. Whatever the method, the important factor is that it should be effective and cost-oriented throughout the marketing chain if losses are to be minimized. Inefficiencies can lead to physical as well as financial losses.

14.3 Implications for labour

All development projects involving technological change affect employment, and projects on the prevention of food losses are no exception. A study in a traditional rice-producing area in Asia has shown that even when food losses were not reduced with the introduction of the pedal thresher and the rice mill, there was considerable labour displacement. In fact, the innovations were introduced because they were labour-saving.

Conservation of labour demand is important because it is essential to show that the extra labour or labour-saving involved in proposed innovations does not occur when there is a prospective surplus of such labour elsewhere for cultivation or processing. A method commonly used to analyse the distribution of labour is a histogram of the labour requirements of an average farming family during one year.

References

A'BROOK, J. Artificial drying of groundnuts: a method for the small farmer. *Trop. Agr.*,
1963   40(3): 241-245. [July]
AMERICAN ASSOCIATION OF CEREAL CHEMISTS. *Post-harvest grain loss assessment methods*,
1978   eds K.L. Harris and C.J. Lindblad. USAID. 193 p.
AMERICAN SOCIETY OF AGRICULTURAL ENGINEERS. *Post-harvest losses in West Africa*, by
1979   G. Yaciuk and R.S. Forrest. Paper 79-5062.
BOLDUC, F.N. & CHUNG DO-SUP. *Development of a natural convection drier for on-farm use*
1978   *in developing countries*. Research Report No. 14. Kansas State University. [December]
FAO. *Manual of fumigation for insect control*, by H.A.U. Monro. FAO Agricultural Study
1969   No. 79. Rome. 381 p.
FAO. *Handling and storage of foodgrains in tropical and subtropical areas*, by D.W. Hall.
1970   FAO Development Paper No. 90. Rome. 350 p.
FAO. *Rice testing methods and equipment*, by F. Gariboldi. FAO Agricultural Services Bul-
1973   letin No. 18. Rome. 55 p. [Mimeo]
FAO. *Rice milling equipment operation and maintenance*, by F. Gariboldi. FAO Agricul-
1974   tural Services Bulletin No. 22. Rome.
FAO. *FAO Action Programme for the Prevention of Food Losses: guidelines and proce-*
1978   *dures*. Publication W. L2783. Rome.
FAO. *Illustrated glossary of rice-processing machines*, by I. Borario and F. Gariboldi. FAO
1979   Agricultural Services Bulletin No. 37. Rome. 37 p. [Mimeo]
FAO. *Manual of procedures to measure the losses occurring during drying, cleaning and*
1980   *handling of grain at farm and village levels*, by D.J. Greig. PFL Unit. Rome. 70 p. [Mimeo]
FAO. *On-farm maize drying and storage in the humid tropics*. FAO Agricultural Services
1980   Bulletin No. 40. Rome. 60 p. [Mimeo]
FAO. *A techno-economic evaluation of rice mills for cooperative and village operations*, by
1980   J. Ramalingam. PFL Programme, Bulog, Indonesia. Rome. 26 p. [Mimeo] [December]
FAO. *Loss assessment*, by G.G.M. Schulten. PFL Unit. Rome. 30 p. [Mimeo].
1982
FAO. *Rodent control in agriculture*, by J.H. Greaves. FAO Plant Production and Protection
1982   Paper No. 40. Rome. 88 p.
FAO. *The assessment of post-harvest losses in Swaziland*, by C.P.E. De Lima. PFL Unit Proj-
1983   ect SWA/002/PFL. Rome. 60 p. [Mimeo]
FAO. *Consultancy report on pest management*, by C.D.F. De Lima. Rome. 53 p.
1983
FAO. *Cooperative marketing development for KUDs and farmer groups*. Project INS/78/
1983   067. Tasikmalaya, West Java, Indonesia. 46 p. [Appendixes]
FAO. *Improvement of fruit and vegetable marketing in China*. Mission Report. Rome. 95 p.
1983   [Appendixes]
FAO. *Social and economic aspects of prevention of food loss activities*, by K.A.P. Stevenson.
1984   Rome. 28 p.
GASGA. *GASGA Seminar on the Appropriate Use of Pesticides for the Control of Stored-*
1981   *product Pests in Developing Countries*. Slough, Berkshire, TDRI. 202 p.
GASGA. *GASGA Workshop on the Larger Grain Borer* Prostephanus truncatas *(Horn)*.
1983   Eschborn, GTZ. 139 p.
GRACEY, A.D. & CALVERLEY, D.J.B. Grain stores for tropical countries. *Trop. Stored*
1979   *Prod. Inf.*, 37: 25-30.
GRACEY, A.D. Construction of new storage facilities: avoidable problems. *Trop. Stored*
1981   *Prod. Inf.*, 41: 13-18.
TETER, N.C. *South East Asia Cooperative Post-Harvest Research and Development Pro-*
1979   *gramme*. 8 vols. Laguna, Phillipines. [November]
UNDP/FAO. *Marketing Development for the Transmigration Settlement Areas*, by P.E.
1983   Percy. Working Document UNDP/FAO Project INS/78/012. Jakarta. 48 p. [Annexes]

WHERE TO PURCHASE FAO PUBLICATIONS LOCALLY

Algeria	Société nationale d'édition et de diffusion, 92, rue Didouche Mourad, Alger.
Argentina	Librería Agropecuaria S.A., Pasteur 743, 1028 Buenos Aires.
Australia	Hunter Publications, 58A Gipps Street, Collingwood, Vic. 3066; Australian Government Publishing Service, Sales and Distribution Branch, Wentworth Ave, Kingston, A.C.T. 2604. Bookshops in Adelaide, Melbourne, Brisbane, Canberra, Perth, Hobart and Sydney.
Austria	Gerold & Co., Graben 31, 1011 Vienna.
Bahrain	United Schools International, PO Box 726, Manama.
Bangladesh	ADAB, House No. 46A, Road No. 6A, Dhanmondi R/A, Dhaka.
Belgium	M. J. De Lannoy, 202, avenue du Roi, 1060 Bruxelles. CCP 000-0808993-13.
Bolivia	Los Amigos del Libro, Perú 3712, Casilla 450, Cochabamba; Mercado 1315, La Paz.
Botswana	Botsalo Books (Pty) Ltd, PO Box 1532, Gaborone.
Brazil	Livraria Mestre Jou, Rua Guaipá 518, São Paulo 05089; Fundação Getulio Vargas, Praia de Botafogo 190, C.P. 9052, Rio de Janeiro; A Nossa Livraria, CLS 104, Bloco C, Lojas 18/19, 70.000 Brasilia, D.F.
Brunei	SST Trading Sdn. Bhd., Bangunan Tekno No. 385, Jln 5/59, PO Box 227, Petaling Jaya, Selangor.
Canada	Renouf Publishing Co. Ltd, 61 Sparks St (Mall), Ottawa, Ont. KIT 56; toll-free calls in Canada: 1-800-267-4164; local and other calls: 613-238-8985.
China	China National Publications Import Corporation, PO Box 88, Beijing.
Congo	Office national des librairies populaires, B.P. 577, Brazzaville.
Costa Rica	Librería, Imprenta y Litografía Lehmann S.A., Apartado 10011, San José.
Cuba	Ediciones Cubanas, Empresa de Comercio Exterior de Publicaciones, Obispo 461, Apartado 605, La Habana.
Cyprus	MAM, PO Box 1722, Nicosia.
Czechoslovakia	ARTIA, Ve Smeckach 30, PO Box 790, 111 27 Prague 1.
Denmark	Munksgaard Export and Subscription Service, 35 Nørre Søgade, DK 1370 Copenhagen K.
Dominican Rep.	Fundación Dominicana de Desarrollo, Casa de las Gárgolas, Mercedes 4, Apartado 857, Zona Postal 1, Santo Domingo.
Ecuador	Su Librería Cía. Ltda., García Moreno 1172 y Mejía, Apartado 2556, Quito.
El Salvador	Librería Cultural Salvadoreña S.A. de C.V., Calle Arce 423, Apartado Postal 2296, San Salvador.
Finland	Akateeminen Kirjakauppa, 1 Keskuskatu, PO Box 128, 00101 Helsinki 10.
France	Editions A. Pedone, 13, rue Soufflot, 75005 Paris.
Germany, F.R.	Alexander Horn Internationale Buchhandlung, Friederichstr. 39, Postfach 3340, 6200 Wiesbaden.
Ghana	Fides Enterprises, PO Box 14129, Accra; Ghana Publishing Corporation, PO Box 3632, Accra.
Greece	G.C. Eleftheroudakis S.A., 4 Nikis Street, Athens (T-126); John Mihalopoulos & Son S.A., 75 Hermou Street, PO Box 73, Thessaloniki.
Guatemala	Distribuciones Culturales y Técnicas «Artemis», 5a. Avenida 12-11, Zona 1, Apartado Postal 2923, Guatemala.
Guinea-Bissau	Conselho Nacional da Cultura, Avenida da Unidade Africana, C.P. 294, Bissau.
Guyana	Guyana National Trading Corporation Ltd, 45-47 Water Street, PO Box 308, Georgetown.
Haiti	Librairie "A la Caravelle", 26, rue Bonne Foi, B.P. 111, Port-au-Prince.
Hong Kong	Swindon Book Co., 13-15 Lock Road, Kowloon.
Hungary	Kultura, PO Box 149, 1389 Budapest 62.
Iceland	Snaebjörn Jónsson and Co. h.f., Hafnarstraeti 9, PO Box 1131, 101 Reykjavik.
India	Oxford Book and Stationery Co., Scindia House, New Delhi 100 001; 17 Park Street, Calcutta 700 016; Oxford Subscription Agency, Institute for Development Education, 1 Anasuya Ave, Kilpauk, Madras 600 010.
Indonesia	P.T. Inti Buku Agung, 13 Kwitang, Jakarta.
Iraq	National House for Publishing, Distributing and Advertising, Jamhuria Street, Baghdad.
Ireland	The Controller, Stationery Office, Dublin 4.
Italy	Distribution and Sales Section, FAO, Via delle Terme di Caracalla, 00100 Rome; Libreria Scientifica Dott. Lucio de Biasio "Aeiou", Via Meravigli 16, 20123 Milan; Libreria Commissionaria Sansoni S.p.A. "Licosa", Via Lamarmora 45, C.P. 552, 50121 Florence.
Japan	Maruzen Company Ltd, PO Box 5050, Tokyo International 100-31.
Kenya	Text Book Centre Ltd, Kijabe Street, PO Box 47540, Nairobi.

WHERE TO PURCHASE FAO PUBLICATIONS LOCALLY

Korea, Rep. of	Eulyoo Publishing Co. Ltd, 46-1 Susong-Dong, Jongro-Gu, PO Box 362, Kwangwha-Mun, Seoul 110.
Kuwait	The Kuwait Bookshops Co. Ltd, PO Box 2942, Safat.
Luxembourg	M. J. De Lannoy, 202, avenue du Roi, 1060 Bruxelles (Belgique).
Malaysia	SST Trading Sdn. Bhd., Bangunan Tekno No. 385, Jln 5/59, PO Box 227, Petaling Jaya, Selangor.
Mauritius	Nalanda Company Limited, 30 Bourbon Street, Port-Louis.
Mexico	Dilitsa S.A., Puebla 182-D, Apartado 24-448, México 06700, D.F.
Morocco	Librairie "Aux Belles Images", 281, avenue Mohammed V, Rabat.
Netherlands	Keesing Boeken V.B., Joan Muyskenweg 22, Postbus 1118, 1000 BC Amsterdam.
New Zealand	Government Printing Office, Government Printing Office Bookshops: 25 Rutland Street; Mail orders: 85 Beach Road, Private Bag, CPO, Auckland; Ward Street, Hamilton; Mulgrave Street (Head Office), Cubacade World Trade Centre, Wellington; 159 Hereford Street, Christchurch; Princes Street, Dunedin.
Nicaragua	Librería Universitaria, Universidad Centroamérica, Apartado 69, Managua.
Nigeria	University Bookshop (Nigeria) Limited, University of Ibadan, Ibadan.
Norway	Johan Grundt Tanum Bokhandel, Karl Johansgate 41-43, PO Box 1177, Sentrum, Oslo 1.
Pakistan	Mirza Book Agency, 65 Shahrah-e-Quaid-e-Azam, PO Box 729, Lahore 3; Sasi Book Store, Zaibunnisa Street, Karachi.
Panama	Distribuidora Lewis S.A., Edificio Dorasol, Calle 25 y Avenida Balboa, Apartado 1634, Panamá 1.
Paraguay	Agencia de Librerías Nizza S.A., Casilla 2596, Eligio Ayala 1073, Asunción.
Peru	Librería Distribuidora «Santa Rosa», Jirón Apurímac 375, Casilla 4937, Lima 1.
Philippines	The Modern Book Company Inc., PO Box 632, Manila.
Poland	Ars Polona, Krakowskie Przedmiescie 7, 00-068 Warsaw.
Portugal	Livraria Bertrand, S.A.R.L., Rua João de Deus, Venda Nova, Apartado 37, 2701 Amadora Codex; Livraria Portugal, Dias y Andrade Ltda., Rua do Carmo 70-74, Apartado 2681, 1117 Lisbonne Codex.
Romania	Ilexim, Str. 13 Dicembrie No. 3-5, Bucharest Sector 1.
Saudi Arabia	The Modern Commercial University Bookshop, PO Box 394, Riyadh.
Singapore	MPH Distributors (S) Pte. Ltd, 71/77 Stamford Road, Singapore 6; Select Books Pte. Ltd. 215 Tanglin Shopping Centre, 19 Tanglin Road, Singapore 1024; SST Trading Sdn. Bhd., Bangunan Tekno No. 385, Jln 5/59, PO Box 227, Petaling Jaya, Selangor.
Somalia	"Samater's", PO Box 936, Mogadishu.
Spain	Mundi-Prensa Libros S.A., Castelló 37, Madrid 1; Librería Agrícola, Fernando VI 2, Madrid 4.
Sri Lanka	M.D. Gunasena & Co. Ltd, 217 Olcott Mawatha, PO Box 246, Colombo 11.
Sudan	University Bookshop, University of Khartoum, PO Box 321, Khartoum.
Suriname	VACO n.v. in Suriname, Domineestraat 26, PO Box 1841, Paramaribo.
Sweden	Books and documents: C.E. Fritzes Kungl. Hovbokhandel, Regeringsgatan 12, PO Box 16356, 103 27 Stockholm. Subscriptions: Vennergren-Williams AB, PO Box 30004, 104 25 Stockholm.
Switzerland	Librairie Payot S.A., Lausanne and Geneva; Buchhandlung und Antiquariat Heinimann & Co., Kirchgasse 17, 8001 Zurich.
Tanzania	Dar-es-Salaam Bookshop, PO Box 9030, Dar-es-Salaam; Bookshop, University of Dar-es-Salaam, PO Box 893, Morogoro.
Thailand	Suksapan Panit, Mansion 9, Rajadamnern Avenue, Bangkok.
Togo	Librairie du Bon Pasteur, B.P. 1164, Lomé.
Tunisia	Société tunisienne de diffusion, 5, avenue de Carthage, Tunis.
Turkey	Kultur Yayinlari Is-Turk Ltd Sti., Ataturk Bulvari No. 191, Kat. 21, Ankara. Bookshops in Istanbul and Izmir.
United Kingdom	Her Majesty's Stationery Office, 49 High Holborn, London WC1V 6HB (callers only); HMSO Publications Centre, Agency Section, 51 Nine Elms Lane, London SW8 5DR (trade and London area mail orders); 13a Castle Street, Edinburgh EH2 3AR; 80 Chichester Street, Belfast BT1 4JY; Brazennose Street, Manchester M60 8AS; 258 Broad Street, Birmingham B1 2HE; Southey House, Wine Street, Bristol BS1 2BQ.
United States of America	UNIPUB, PO Box 1222, Ann Arbor, MI 48106.
Uruguay	Librería Agropecuaria S.R.L., Alzaibar 1328, c.c. 1755, Montevideo.
Yugoslavia	Jugoslovenska Knjiga, Trg. Republike 5/8, PO Box 36, 11001 Belgrade; Cankarjeva Zalozba, PO Box 201-IV, 61001 Ljubljana.
Zambia	Kingstons (Zambia) Ltd, Kingstons Building, President Avenue, PO Box 139, Ndola.
Other countries	Requests from countries where sales agents have not yet been appointed may be sent to: Distribution and Sales Section, FAO, Via delle Terme di Caracalla, 00100 Rome, Italy.

Foto-Tipo-lito SAGRAF - Napoli